Evolution Is A Myth

GILBERT C WARD

Copyright © 2012 Gilbert C. Ward

Available at Amazon.com and other retailers

All rights reserved. No part of this book may be used or reproduced by any means, graphic, electronic, or mechanical, including photocopy, recording, taping or by any information storage retrieval system without the written permission of the publisher except in the case of brief quotations embodied in critical articles and review.

Refuting Evolution by Jonathan Sarfati, Ph.D
First printing May 1999
Sixth printing September 1999
Copyright 1999 by Answers in Genesis
Used by permission by publishers:
Master Books Inc., P. O. Box 727
Green Forest, AR 72638

Archaeological Supplement by G. Frederick Owen D. D Ed. D
Copyright © 1964

This information was taken from the Thompson Chain-Reference Bible with permission of Kirkbridge Bible Company. 54th Reprint.

Quotes from Francis Collins: Pew Research Center's Forum on Religion and Public Transcript from May 2009. The entire presentation can be found:
HYPERLINK "mailto:tmiller@pewforum.org" tmiller@pewforum.org or
HYPERLINK "http://Pewforum.org" http://Pewforum.org

All scripture quotation from the King James Version of the Bible.

Because of the dynamic nature of the Internet, any Web addresses or links contained in this book may have changed since publication and may no longer be valid. The views expressed in this book are solely those of the author and do not necessarily reflect the views of the publisher, and the publisher hereby disclaims any responsibility for them.

ISBN: 1469913097
ISBN 13: 9781469913094

Library of Congress Control Number: 2010927356
CreateSpace, North Charleston, SC

Printed in the United States of America

CONTENTS

Introduction vii

Chapter 1 The Consciousness of God 1

Chapter 2 The Consciousness of Science 13

Chapter 3 The Consciousness of History 59

Chapter 4 The Consciousness of the Bible 81

Chapter 5 The Consciousness of the Holy Spirit 93

Chapter 6 The Consciousness of the Time 121

Documentation 133

INTRODUCTION

The purpose of this book is to present evidence that proves evolution is not the scientific project it claims to be; but is a theory that is based upon assumptions and not upon facts. Let me put a question to you! Can you name one single scientific fact that you know that proves evolution? Stop and think!

Fifteen years ago I began to write articles against evolution and felt sure some PhD would pin my ears back; but I was never challenged.

The variety of the species, such as finches, cats, and dogs do not prove evolution. Evolution would be proved if there was evidence showing the links between the species; one to the other; but in spite of having billions of fossils to study, none support evolution. Each fossil can be identified to the species it belongs and the living species look the same as the fossil. There has been no change! Our study will include facts from several fields of science; Anthropology, Ethnology, Biology, to name a few.

Because evolution is based upon assumptions rather than evidence, it cannot allow the competition of Intelligent

Design or any opposing thought to be presented. That is why the Bible had to be removed from public schools. Did you know the evolutionist have been instructed not to debate the Creationists because they always loose.

We have witnessed the deterioration and decay of America in direct ratio with the removal of Christian principles, the Bible and prayer from our Society. We are much like those spoken of in the Bible in Isaiah 5: 20, 21:

> "Woe unto them that call evil good, and good evil; that put darkness for light, and light for darkness; that put bitter for sweet and sweet for bitter! Woe unto them that are wise in their own eyes, and prudent in their own sight."

We pray that this book will be a blessing and a challenge to the reader. God would also want us to know what He says in Ecclesiastes 11: 9:

> "Rejoice, O young man in your youth; and let your heart cheer you in the days of your youth, walk in the ways of your heart, and in the sight of your eyes; but know, that for all these things God will bring you into judgment."

God has a plan for your life and it can be found in His Word; the Bible!

❊ ❊ ❊

CHAPTER 1

THE CONSCIOUSNESS OF GOD

It makes no difference how many times the atheist proclaims there is no God, he cannot convince himself or others, that it is true! If there is no God, why is so much money, time, and energy spent trying to prove that there is no God? If there is no God, then there would be no need for Communism, Marxism, or Humanism; because part of their teaching is that there is no God.

The very fact that so much effort is being put forth to remove God and to shut Him out of our society proves that He does exist! Evolution from its basic theory would not produce a god; it would not be in the mind of the animal to do so. Religion would have to develop slowly as man developed; but this is not what we find!

Every place where man has been found, he had religion. The religion was well developed; it had a belief in a supreme being and an expression of that belief. This has not changed throughout the history of mankind. There is implanted within every human heart the knowledge that there is a God and a desire to know Him. It is implanted in our inner being when we were born.

I would like to ask the reader a personal question. During your life has there not been a time when you thought about God; and maybe, even prayed to Him? I believe your have! The desire to know God is found in the scratches on the sides of caves; and is expressed in writings, music, art, and the magnificent structures around the world that were created and built for the sole purpose of knowing God and worshiping Him.

In the study of Ethnology we learn that every place where man has been found; no matter how simple or how elaborate his culture, there was evidence that he had a belief in a supernatural and had an expression of that belief. It did not matter if it was a piece of carved wood or a huge structure that man has made; man worshiped a god. Man did not evolve to religion; religion was there when man began!

This is also true in regards to clothing, cooking utensils, fire, weapons, music, and language. These were always present wherever man was found. If there is no God; then how is it that throughout history man has devoted much of his time in seeking God and worshiping Him!

Recently we were in New York City and were able to visit the magnificent Cathedral of St. John the Divine. It is advertised as the largest Cathedral in the world. The doors entering the building are constructed in bronze and each weighs 3 tons and measures 18-feet high and 6 feet wide. They are comprised of 60 panels showing in bas-relief scenes from the Old and New Testament. The forty foot diameter Great Rose window in the West Facade is the largest in the United States and contains 10,000 pieces of glass. The length of the building is 601 feet and the Statue of Liberty, minus her pedestal, would stand without any problem under the dome in this immense structure.

Note: A part of the building's length is class rooms, offices, and meeting rooms.

There are eight massive granite columns that surround the High Altar that were quarried on the Island of Vinalhaven, Maine. Each is 55 feet tall, 6 feet in diameter, and weighs 130 tons. One standing in this structure, cannot help but be overwhelmed by what he is seeing. I believe this is good evidence that there were some people back in 1872, who thought it worthwhile to plan, build, and pay for this structure so they could use it to worship God. I have no knowledge of the cost; but as I stood there that day I was reminded of what my professor of Anthropology said in class one day:

"If I have Christ in my heart; I can worship God in a chicken coop!"

This giant structure, in much smaller sizes, is found all over the world representing millions of people that come together each week to worship their God. This has been going on since the beginning of time. A good example of the need to worship is also found with the early Egyptians; the Mayans and Aztecs of Mexico; and the Incas of South America. We also should mention the religions of Hinduism, Buddhism, Islam as well as the different denominations throughout the world that worship their supreme being. As I said before, the evidence supports the fact that religion has always been an important part of man's culture, from the beginning.

It is interesting to note that the evidence of man's civilization only goes back some 6000 years...; as recorded in the Bible. Before that, we only have the artist's conception of imaginary creatures to fill man's past; many of which turned out to be frauds; such as the Pithecanthropus Erectus.

Religion has many forms and extremes. Many believed it took human sacrifices to atone for their sins; like the Aztecs who sacrificed thousands of victims each year for that purpose. Some religions have strange rituals that we in modern times find hard to believe. For example, in sections of India the cobra snake is worshiped.

A missionary, back from India told me this story. One day he said, there was a lot of commotion at the mission compound. A mother and father rushed into the clinic holding their child who did not look well. "He was bitten by a cobra", they shouted. The only antidote was several miles away at the hospital and shouting orders for the parents, with the child, to get into the Jeep, we rushed there.

After sometime, the doctors came out and announced that the child was dead..., and then added the solemn news, that the child had been dead for three days.

We then learned that the snake that bit the child was a household pet and part of the parent's religion. When the child was bitten, they spent two days with their witch doctor seeking healing by their gods and when that didn't happen; as a last resort, they brought the child to the missionaries. It is interesting that in the Bible it speaks of the depravity of the human race deteriorating to the point where they would worship "creeping things". (Romans 1: 22, 23)

Dr. Francis S. Collins, M. D., PhD, a former director of the Human Genome Project, related his story in the Pew forum, of how he came to believe in God. He said he was an atheist because he assumed there wasn't enough evidence for the idea that a God existed.., and atheism was a convenient answer.

His work took him often to the bedside of those who were dying and he began to notice that those who died with faith did so quietly and with peace.

This caused him to wonder how he would face death when it came. He said, being a scientist, he expressed this thinking:

> "Obvious evidence is; we see it all around us, there is something instead of nothing. There is no reason there should be anything at all...unless there is a God. Why should there be any structure, any continuity of reason; any exactness of life or science; without God."

> "The exactness of mathematics, the laws of Physics, the Properties of matter and energy and the law of Gravity; the equation of electromagnetism; all work out to be true! Can the universe have a beginning out of nothing? Nature cannot create nature! For if all of this is true; then we have a God that works outside of time and space!"

He continued:

> "The Big Bang, the fact that the universe had a beginning out of nothingness, as far as we can tell; from this unimaginable singularity, the universe came into being and has been flying apart ever since...that cries out for some explanation. Since we have not observed nature create it self; where did it come from? That leaves; there must be a creator!" 1.

The words of Dr. Collins ring true! There is no way a person with an honest, sincere heart, seeking to know the truth, can look and study the world around him, without concluding there is a God. The night sky with its million of stars expanding beyond the most powerful telescopes cannot be explained by the "Big Bang". The Psalmist wrote in Palms 19:

> "The heavens declare the glory of God."

If we are honest, we must accept that religion has played an important part, from the beginning of time to our modern day; it is not a come lately thing. There is this inner desire to know God and to worship Him; it is within us! We will never find inner satisfaction in life until we have yielded our hearts to God.

II Corinthians 5: 17 states:

> "If any man be in Christ, he is a new creation old things have passed away and all things become new."

Since the evidence of religion has been found as part of man's culture, from the beginning of time; it has continued throughout history. As we stated before; no place or time; has man ever been found that he didn't believe in a supreme being and had an expression of it. If evolution were true, then at some time the evidence of religion would not be found; it would come at a later time as man developed. But this is not the case. As we shall see in Chapter III; the history of civilizations is very much involved with religion.

May I ask the reader: What is your knowledge of God?

From the Bible we are told we are made in the image of God. It speaks of God's eyes; His ears; His breath; His arms; and His emotional characteristics; He can become angry!

Maybe the Bible speaks of Him this way, figuratively, so we can understand Him better.

The Bible also speaks of God's compassion, His faithfulness, gentleness, goodness, longsuffering, His love, His majesty, mercy, His nearness, protection, His sovereignty, His righteousness, and His holiness. These are some of the attributes of God. He also is the One who flung the stars in space....and wants us to call Him our Heavenly Father.

How can I explain God to the reader? Jesus is speaking to the scribes and Pharisees in the 15th chapter of the Book of Luke and gives them an example of what God is like. He told them that God was like a shepherd who had a hundred sheep and one night one was missing. He saw to it that the ninety nine sheep were safe in the fold, and then started out to find the lost one.

When he found the lost sheep, He laid it across His shoulders and carried it back to safety and there was great rejoicing because the sheep that was lost was found. This parable of Jesus has been put to music, it's called: the Ninety and Nine:

> There were ninety and nine that safely lay,
>> in the shelter of the fold,
>
> But one was out on the hills away,
>> Far off from the gates of gold.
>
> A way on the mountains wild and bare,
>> Away from the tender Shepherd's care,
>
> A way from the tender Shepherd's care.
>
> Lord thou hast here Thy ninety and nine;
>> Are they not enough for Thee?
>
> But the Shepherd made answer: "This of mine

has wandered away from me,
And although the road be rough and steep
    I go to the desert to find my sheep,
I go to the desert, to find my sheep."

But none of the ransomed ever knew
    How deep were the waters crossed?
Nor how dark was the night that the Lord
    passed thro
Ere He found His sheep that was lost.
    Out in the desert He heard its cry-
Sick and helpless and ready to die;
    Sick and helpless and ready to die.

Lord, whence are those blood-drops all the way,
    That mark out the mountain track?
"They were shed for the one who had gone astray,
    Ere the Shepherd could bring him back."
Lord, whence are Thy hands so rent and torn?
    They're pierced tonight by many a thorn;
They're pierced tonight by many a thorn.

But all thro' the mountains thunder-riv'n,
    up from the rocky steep,
There arose a glad cry to the gate of heaven,
    "Rejoice! I have found my sheep!"
And the angels echoed around the throne,

> "Rejoice, for the Lord brings back His own!
> Rejoice, for the Lord brings back His own!" 2.

This song expresses the love God has for each of us. That He is searching for us to come to Him. That He paid a great price, the crucifixion of His Son, Jesus Christ, to show His love and desire to embrace us.

Could you be one of those lost sheep?

Also in the same chapter of Luke 15 is the parable of the disobedient son who would not listen to his father's instructions and chose to go out from his home environment to live in the waste of society.

It says in a far country he spent his money foolishly and when it was gone, he found food in the garbage dump. That's when he thought of home and hoped he could get a job working for his father and so he made that regrettable, embarrassing journey back to his family. To his surprise, when he was still a far distance, his father saw him and came running to meet him. The son did not have a chance to ask his father for a job; his father received him with open arms and ordered a welcome home party because:

> "This, my son was dead and is alive again."

The first story is an example of God seeking, searching, and finding...; the second; is that of the Father forgiving and receiving. I do not know the reader's position; but from this is the opportunity, if needed.., to respond to the One who is seeking you and is ready to forgive you. This is what God is like!

God's interest in our personal lives is sometimes difficult to accept; but only because we are a little bit standoffish about turning everything in our lives over to Him. That would be like someone always telling us what to do! But God is not that intrusive. We still have our own will; but the blessings come as we learn to surrender our will onto His.

In proverbs 3: 5, 6, reads: "Trust in the Lord with all your heart and lean not to your own understanding. In all your ways acknowledge Him and He shall direct you path."

"God is a merciful God; His mercy endurith for ever." Many of us have grown up, so to speak, in the church. It was a ritual we went through every Sunday with our families and when we came of age, our concepts of God, church, and religion, faded from our lives as we matured to outside influences. They were more exciting than the church and maybe that is the experience of the reader.

I was just about like that, when I visited a private home on the North side of Chicago, many years ago. I still remember the address; it was 1919 W. Leland Ave.; and there, for the first time, I became aware that God wanted a personal relationship with me! Not a group relationship, as in a church membership; but a one on one experience.

At the meetings I heard the testimonies of Christians sharing their experiences with Christ; how He had changed their lives, filled them with joy, and was leading them each day. I realized that this was what I was looking for, a personal acceptance of Jesus Christ as Lord of my life! I was seventeen at the time and later served in WW II; so that gives you an idea of my age! I have walked with the Lord all these years and I've never regretted it!

No person has ever fully comprehended the love that God has for us. We have tied to express it in words and music; and no matter how moving the words may be, or how beautiful the music; it does not touch the edge of His love. The words of this song, comes close; perhaps you know them?

THE CONSCIOUSNESS OF GOD

The love of God is greater far,
> than tongue or pen can ever tell.

It goes beyond the highest star,
> and reaches to the lowest hell.

The guilty pair, bow down with care,
> God gave His Son to win,

His erring child, He reconciled,
> and pardoned from his sin.

When hoary times shall pass away,
> And earthly thrones and kingdoms fall,

When men who here refuse to pray,
> On rocks and hills and mountains call.

God's love so sure, shall still endure
> All measureless and strong,

Redeeming grace to Adam's race-
> The saints and angels song.

Could we with ink the ocean fill?
> And were the skies of parchment made,

Were every stalk on earth a quill?
> And every man a scribe by trade.

To write the love of God above,
> Would drain the ocean dry.

And could the scroll, contain the whole,
> Though stretched from sky to sky.

> Oh love of God, how rich and pure!
>
> > How measureless and strong.
>
> It shall forever more endure,
>
> > The saints and angels' song. 3.

The way we come to know God is through Jesus Christ His Son. Jesus said: "I am the way; the truth; and the life; no man comes to the Father except by me!" John 14: 6.

Sometime in our life, we will come to the realization we need God; whether it is out of desperation, or like Dr. Collins, who sought answers to questions about the faith of Christians when they faced death, we feel the need of knowing Him. God will speak to our heart, our inner self..., and we must respond.

You say there is no God! Why not put Him to a test right now? Just say: God...; if you're real....; show yourself to me.

❀ ❀ ❀

CHAPTER II

THE CONSCIOUSNESS OF SCIENCE

Much like Dr. Collins experience in his life, many people give little thought that science and God are related. Doesn't it seem reasonable, that if there is a God and He created the Universe, that science, which He created, would be in harmony with Him? However, we are bombarded so many times from the media that the world is millions of years old, that we begin to believe it must be true. The evidence of civilization only goes back six thousand years. If the world is millions of years old, like they say, wouldn't there be signs of civilizations before six thousand years?

Evolution is presented as a scientific fact, and the Bible is referred to as a book of myths and children stories. Actually, it is the other way around. The Bible is supported by science and evolution is full of myths and children stories.

Let me make a statement here:

"There is not one single scientific fact that proves evolution! The Bible is the Word of God and it is supported by science!"

Evolution, much against what the evolutionist would have us believe, is not the name of a science! It is only the name of an idea based upon assumptions that fits well into the mind of those who wish there was no God. Science is founded upon facts that can be proven by experimentation or visual evidence.

ARCHAEOLOGY

Archaeology is the name of a science and it gives us visual proof that the Bible is true! It is the study of the remains of ancient civilizations and bases its conclusions on the evidences it finds in those excavations! It does not work on a theory that each discovery has to fit into; but stands alone for what it is! In their work they have uncovered and translated writings that reveal the names of people; cities; kingdoms; and events that took place as far back as 6000 years.

Every archaeological excavation in the land of the Bible, has documented the truth of the Biblical record! The early history of civilization does not speak of the physical development of man from the simple to the complex; it speaks of intelligent people living on different levels of society in different parts of the world throughout our known history! Egypt was a flourishing country with pyramids that have stood since 3000 B.C.! Mesopotamia, the area around the Tigris and Euphrates Rivers, also had well developed civilizations.

Many years ago, a great number of scientists, representing different countries, met at the Oriental Institute at the University of Chicago to determine once and for all, where civilization first began and it was concluded that the cradle of civilization, the first evidence of human life, started in the area of Mesopotamia, which is the location of the Biblical Garden of Eden.

Today, the evolutionist are trying to change this location to Africa based upon the finding of Lucy and Ardi, two fossils images they are trying to use as a connection between man and ape. However, these two fossils do not constitute a civilization and hardly qualify to be the cradle or beginning of it.

In the study of Archaeology, none of the findings at any excavation ever contradicted the historical records of the Bible; indeed, their findings have confirmed the Bible! Many times the archaeologists refer to the Bible in helping them locate the land marks they need to find for their next excavation. Here are some of the findings of Archaeology.

In 1842, Emile Botta was excavating an area ten miles north of Nineveh near the Tigris River in Assyria, better known as Iraq. Soon into the dig they found evidence of a past civilization. The more ground they uncovered, the more they realized that this was a find that would amaze the world. It was huge in size and as they proceeded a city was revealed that covered 741 acres. It had a palace that covers 25 acres and was surrounded by a wall that range from 9 ft. to 25 ft. thick. The palace was built upon a brick-terraced platform that raised the palace forty-five feet above the rest of the city; this making the palace visible for miles around as one approached it.

The palace had spacious domestic quarters with a luxurious harem, large reception halls and living quarters. The walls were adorned with inscriptions and reliefs that displayed their gods, and giant winged bulls weighing in at ten to thirty tons each, guarded the halls and doorways. The inner floors were laid with tile or tamped clay and were covered with fine rugs. The outer courts and open spaces were paved with highly colored tile or marble blocks, and there, on the walls of the palace, were the seemingly endless successions of sculptured pictures portraying in grand detail the daily life, pleasures, appearances, customs, religion, and history of the Assyrians. This was a monstrous find!

The palace and city belonged to King Sargon who reigned over Assyria for fifteen years. There is no doubt about his name because it is inscribed on the walls of the city and palace. One marble relief shows an Assyrian nobleman being greeted by King Sargon.

However; the odd thing to note about this king is that there were no historical writings or records to show this king ever existed! That is until Paul Emile Botta found the palace and city back in 1842. But what is still even more unusual is that this king's name is recorded in the Bible! 4.

In the Book of Isaiah chapter 20 verse 1, it reads:

> "In the year of Tartan came unto Ashdod,
> when Sargon, the King of Assyria sent him,
> and fought against Ashdod and took it".

For years, this King's name was a point of criticism of the Bible because his name is only recorded once in the scriptures. They argued that the author needed a name to fit into his story and made up the name, King Sargon. However; today, we know a great deal about this ancient King and his life because the evidence is on display in many museums around the world.

The question is: 'How did this king's name ever get into the Bible?' The book of Isaiah was written during the area of

740 and 705 B. C. No one at a later date could go back and enter King Sargon's name into the ancient scrolls of the Bible! The only way King Sargon's name could be in the Bible is that the author, who wrote the book of Isaiah, was living at that time when these events took place; therefore, this history of King Sargon was recorded by an eye witness!

Here is one of the translations that were carved on one of the walls in Sargon's palace:

> "In the first year of my reign the city of Samaria I besieged and captured. 27,290 people from its midst I carried captive. 50 chariots I took there as an addition to my royal force. I returned and made more than formerly to dwell. My officers over them as governors I appointed. Tribute and taxes I imposed upon them after the Assyrian manner." 4.

The siege of Samaria was started by King Shalmaneser of Assyria; but was completed by King Sargon who replaced Shalmaneser. The account is recorded in the Bible in II Kings the 17th chapter.

It speaks of the King of Samaria, Hoshea; dealing falsely with King Shahmaneser, the king of Assyria and in the disagreement caused Shalmaneser to attack Samaria. The war went on for three years and somehow, the record is not clear; king Shalmaneser was replaced by King Sargon, who won the conflict as found in Isaiah 20: 1.

Ashdod, in the verse, refers to a city in Samaria and that city is still there today located along the coast of the Mediterranean Sea in Israel.

Here we have two confirmations about the accuracy of the Bible. One, that the name of King Sargon is documented by the finding of his palace, and two; the inscription on King

Sargon's wall proves the Biblical account of Israel's defeat and their enslavement in Assyria.

Let's look at another example from the field of archaeology that shows the accuracy of the Bible. In 1887 a peasant woman digging for fertile soil at the mound of Amarna; discovered a cuneiform tablet. This uncovered the royal archives of King Amenhotep III, and his son, Amenhotep IV. They were rulers of Egypt at the time these tablets were written. During the excavation, additional cuneiform tablets were found totaling more than 350, and they are all dated around 1358-1350 B. C., the times these kings lived. They are known as the Tell el-Amarna tablets and can be seen in the National Museums of England, France, Egypt, and Germany.

These tablets were communication dispatches between the king of Egypt and the governors and officers located in different cities and fortresses in the Palestine area. Some of the tablets were of a personal nature such as; the daughter that the king promised had not yet arrived, and money owed had not been paid. Others simply spoke of natural experiences of life and the community.

Among the tablets were those that spoke to the Egyptian king about an invading force coming into the area and assuring him that there would be no problems. But later a request for help was made, and later still, a more frantic cry for help. This is the translation:

> "Let my Lord the King, the sun of heaven, take heed unto his land, for the Khabiri are mighty against us; and let the king, my lord, stretch out his hand unto me and let him deliver me from their hands, so they may not make an end of us." 4.

The king wrote back advising him to combine forces with the other rulers of the area; and he received this message back:

> "To the king my lord, say. Thus saith Abdi-hebda thy servant. At the feet of the king, my Lord, seven times and seven times I prostrate myself...the whole land of the king has revolted. There is not one governor that is loyal to the king, all have rebelled. The Kabiri are capturing the fortresses of the King, All have rebelled, all have rebelled. May the King harken to unto Abdi-Jeba and send troops, for if no troops come this year the whole territory of my lord the King will be lost. The Habiru are capturing the fortresses of the King. May the King care for his land. The Habiru are taking the cities of the King. If there are no archers this year, then let the King send a deputy that he may take me to himself together with my brothers and we die with the King, our lord." 5.

All of this has little meaning to us until we understand the meaning of the words: "Khabiri" and "Habiru". These were Egyptian words for: "Hebrew"!

These tablets are giving the account of the Jewish people, the Hebrew people, who were released as slaves from Egypt and were advancing to conquer the land that God had promised them. You remember the movie: "The Ten Commandments", with Charlton Heston, well..., the Tel Amarna Tablets confirms that this really happened! The Jewish people really were slaves in Egypt and a man named Moses did lead them out as recorded in the Bible.

The Amarna Tablets fit perfectly into the time frame of 1400 B. C., when Israel was moving into the Promised Land. (Fairy tales, you say!)

Another case: Dothan is a city that existed back in 3000-2000 B.C. It is the place in the Bible where Joseph was placed in a dry well by his brothers and later sold as a slave into Egypt. Genesis 37: 17-28.

Dothan is also where Elisha, the prophet, when seeing the attacking armies approaching declared: "Fear not; for they that be with us are more than be with them." II Kings 6: 13-23.

This lost city was found in the 1950's by archaeologists proving again that the Bible's historical records are trust worthy. This information is not coincidental; the names of people, places, and events in the Bible are again and again proven that they existed...; by the science of archaeology. 6.

If the atheist wishes to keep his faith, he should never travel to Israel because in every turn of the road, every place one visits, the evidence confirms the truth of the Bible. The Bible has always been under attack by those who have never studied it and have a personal agenda they want to promote. Depending upon the exposure the reader has had; this evidence may be difficult to accept; but the Bible has endured and will continue because God has promised it would! Remember in this study, we are to examine the evidence, not assumptions..., to learn the truth!

In G. Frederick Owens's writings on the subject of Archaeological findings, he wrote:

> "God kept two copies of the historical records of His special dealing with, and revelation to man. One was the Bible which had been written on parchment, and by great efforts, placed into the hands of man; and the other, was written in the ruined

remains and strange languages of these lands from whence the Bible came." 7.

People, names, and places in the Bible, that seem insignificant, when reading the Bible, are confirmed by Archaeology. Aphek was an important outpost back in the days of Pharaoh Thutmosis III. It was excavated and there was found a tablet that referred to the king of Aphek! He is listed along with other kings that ruled in other countries surrounding Israel. This list of names is found in the Book of Joshua chapter 12; verses 7-24.

Aphek was one of the Kings Joshua fought against in Israel's conquest of the Promised Land. King Jehu was the ruler over Israel at the time and ruled for twenty eight years. In the excavation of Calah, later called Nimrud, there was found a great deal of information about the history that took place during that time of 885-782 B.C.

One of the important finds was the "Black Obelisk" which was set up by Shalmaneser III. The monument is made of black marble and is six feet, six inches tall with a tapered top. It has twenty small bas-reliefs, five on each side, showing the officials from different countries bringing tribute to the king. There are 210 lines of cuneiform inscriptions which tell the story of the monarch's achievements. At the end of one of the inscriptions are these words:

> "I then marched as far as the mountains of Ba'lira'si, by the sea-side, and erected there a pillar, stela, with my image as King. At that time I received the tribute of the inhabitants of Tyre, Sideon, and of Jehu, son of Omri". 8.

Here etched in history is the name of this Jewish King, Jehu, who at one time ruled Israel. One might think what is so important about this? It is to show that the Bible is accurate in small details as well as large details. This evidence could not be planted; reconstructed; or assumed like the theory of evolution. The archaeologist has no agenda to prove a theory; he works simply to reveal the evidence he finds as it relates to history. This evidence cannot be denied because in every detail that is found, related to the Bible, the evidence supports the Biblical record.

Therefore, when public leaders make remarks insulting the Bible and those who believe in it; they are doing so out of ignorance of scientific proof. These people can be easily identified because they are the same ones with degraded morals that seek the demise of our Constitution and America.

With the Bible, we are not dealing with hearsay, or assumptions, or myths, because every little detail and every large detail in the Bible has been, or is being substantiated by archaeological evidence. The Bible is the foundation that America was founded upon. Our moral values, our laws, our form of government were all based upon the freedom which comes from the Bible. What we have seen and are seeing in America today, is the planned destruction of every influence the Bible and God has on our society!

This destruction is based upon lies propagated in our schools, some of our churches, and by the media, and politicians. In II Peter 2: 1-3, It speaks of false teachers sneaking in among us who will deny God and mislead the people; even denying that Jesus Christ as Lord!

In 1947, at Ain Fashkha, about seven miles south of Jericho and one mile west of the Dead Sea; some wandering Arabian nomads, carrying goods from the Jordan Valley to Bethlehem, were searching for a lost goat and came upon a cave in which they found a number of crushed jars which contained scrolls. Later, other scrolls were found in the same vicinity and they became known as the Dead Sea Scrolls. Perhaps you have heard of them? 8.

THE CONSCIOUSNESS OF SCIENCE

One of the scrolls was identified as a copy of the book of Isaiah that was originally written around 760 B. C. That's 2, 772 years ago, do you understand!! It was written in ancient Hebrew on parchment that was sewn together in sheets 10 X 15 inches and together measured 24 feet in length. From the history of that location and the people that lived there, the copy would have been written around 200 B.C. What is so amazing about this find is that its content is exactly as what we find in the book of Isaiah today. So here is another confirmation of God's promise:

> "Heaven and earth shall pass away; but my words will never pass away."

In the Kidron Valley, near the southeast corner of the Old Jerusalem City Wall, was found a strange looking monument dating around 100 B.C. It has inscriptions naming it as Absalom's Pillar; known also as Absalom's tomb. It is forty-seven feet high and the lower base is twenty feet square. Absalom was the son of King David and in II Samuel 18: 18, is the reference to the building of this monument:

> "Now Absalom in his lifetime had taken and reared up for himself a pillar, which in the King's dale: for he said, I have no son to keep my name remembrance: and he called the pillar after his name." 9

Here again is archaeological evidence that supports the historical accuracy of the Bible, even in small details. For additional information on this subjects visit your local Christian Book Store. There are many books written on these subjects that confirm the truth of the Bible. Answers in Genesis is also another very good source for scientific information about the Bible.

❋ ❋ ❋

ANTHROPOLOGY

Anthropology is the name of a science and it is the study of man from a physical point of view. This is a difficult study for the evolutionist because for over two hundred years they have been looking for some missing link to support their theory and they have found none. In dealing with evolution, science text books have to be revised every ten years because new evidence requires them to make adjustments in their theory.., such as the single cell to the Big Bang!

If evolution were true and the earth billions of years old, as they claim, there should be millions of connecting links between the species in the fossil record and the stratum of the earth; but this we do not find!

In considering evolution, we must accept first that there was nothing! At one time we were taught that life began as a one cell animal that emerged from some form of pond and it evolved over many millions of years, step by step, to where we are today.

Since the evolutionary process had to start simple, it had to evolve toward the complex, or to a more intelligent form. This myth was really blown apart a few years ago when it was discovered that the single cell was not simple, but very complex. This changed the bases of evolution; the simple to the complex, and so the theory was changed to the Big Bang!

They do not know when the Big Bang took place, where the material came from, or what caused the Big Bang. This is not science; this is theory, better known as assumptions.

From this nothing, the Big Bang, an explosion took place that formed the sun, moon, planets, stars, plants, animals and man. (Now that's a fairy tale, if you can believe it.)

If you visit the Smithsonian Museum of Natural History in Washington D. C., and watch the DVD on evolution you will learn that since the one cell animal didn't work out, we now have evolved from a rat. What is strange in the DVD is that fully developed dinosaurs existed at the same time the rat existed. There was no showing the rat evolving into another species. They also do not give any evidence as to where the rat came from; so it is also based upon assumptions. Personally, if I had my choice, I would much rather think of myself as coming from a one cell animal, than from a rat!

Each year, thousands of school children pass through the Smithsonian and are influenced to believe that science proves evolution to be true. But it is all based upon assumptions. There are no interlinks between the species to prove that we came from a lower form of animal life. What do the leading scientists have to say about this problem of no connecting links?

D.M.S. Watson, a noted biologist, whose book: "Adaptation" states:

> "Evolution is a theory; universally accepted not because it can be proven by logically coherent evidence to be true, but because the only alternative, specially, special creation is clearly incredible. 10.

Notice in this statement the author states that the theory is accepted not because it can be proven; but because it's better that believing in Creation.

In his book: "How the World Works" author, Boyce Rensberger wrote:

> "At this point, it is necessary to reveal a little inside information about how scientist work, something the text books don't usually tell you. The fact is that scientists are not really objective and dispassionate

in their work as they would like you to think. Most scientists first get their ideas about how the world works, not through rigorously logical processes, but through hunches and wild guesses.

"As individuals, they often come to believe something to be true long, long before they assemble the hard evidence that will convince somebody else that it is. Motivated by faith in his own ideas and a desire for acceptance by his peers, a scientist will labor for years knowing in his heart that his theory is correct but devising experience after experience, whose results he hopes will support his position." 11.

Please notice; it is not just the physical species of life that evolutionist must answer to; where is the evidence that supports the changing of a pear tree to a pine tree, or strawberries to grapes; or plants at all! From time to time; great announcements are made, claiming the discovery of some ape like creature that is the missing link, but in time the evidence proves false. They are so eager to find something positive; that they overlook some of the facts.

I remember reading about a scientist in the late 30's who proclaimed he had created life. This caused great excitement in the science world and the day came when he would demonstrate the process of creating life. This was held in Chicago. A large crowd was present at this auditorium and on the platform was the scientist, a table with boiling water, and an assistant. To begin the experiment the scientist took a twig from a box and was about to place it in the cooled water when Dr. Harry Remmer spoke up and asked: "Is the twig sterilized?" the scientist answered: "No, if I do that, the experiment will not work".

The evolutionist would ask us to believe that life came about by non living chemicals; but they cannot say where the chemicals came from. Would you be surprised if I told you there were no pre-historic men; no ape-like persons; that they are myths created by anxious atheists looking for proof that there is no God! Here are the names of some of these myths: The Nebraska Man; the Pithecanthropus Erectus; the Piltdown man, the Neanderthal; and the late Lucy...; and now we must add "Ardi".

The Nebraska Man was created out of a single tooth that Clarence Darrow used to confuse the court and William Jennings Bryon in the famous Scopes Trial. The tooth was presented as belonging to a prehistoric ape like man; but later was proved to be the tooth of an extinct pig that at one time roamed the West. 12.

The Java man, also known as Pithecanthropus Erectus, was found by Dr. Eugene Dubois, a Dutch physician who was serving in the Dutch Army in the Dutch East Indies. He was an avid believer in evolution and spent his spare time looking for evidence that would support that theory. In 1891, on the island of Java, he found a portion of a skullcap and three teeth. A year later he went back and found a thigh-bone, which was about fifty feet away from where he found the skull cap and teeth. A true scientist would not accept that the thigh-bone and skullcap and three teeth were related.

Dubois returned to Holland and announced to the world that he had found the missing link between man and ape. The world accepted it as positive proof of evolution. The news spread around the world in text books, magazines, drawings, and images that portrayed the Java man as a creature that was part ape and part man. It was named: Pithecanthropus Erectus.

However; no one was allowed to see or examine the evidence Dr. Dubois had found because he kept the evidence locked in a safe for thirty years. When the items were finally inspected, it was concluded that two of the teeth were those of an orangutan, the other was human; the leg bone was human; and the part of the skullcap was too small to be that of a man and it was concluded that it was all a fraud. 13.

Of course, there was the announcement to the whole world to that effect, and many apologies were made from the scientific community, and all misleading images and pictures were destroyed....you would think!

But that is not the case. Unless it has been taken down since I was there a few years ago; the Pittsburg Museum of Natural History, in their section on early history, still had on display, a picture of an artist concept of the apelike figure of Pithecanthropus Erectus. Imagine the impact this makes upon children as they pass through that section of the museum on their school tour.

In the March, 2010 issue of the Smithsonian Magazine there is an article: Discovering our Ancestors, and in the article the fossil of the Java man, the Pithecanthropus Erectus is used as evidence as one of our ancestors, even though it was proven to be a fraud.

Another myth from the evolutionist is the Piltdown Man, which consisted of fragments of a skull and a jawbone that were collected in 1912 from the gravel pit at Piltdown, near East Sussex, England. The fragments were thought, by many experts, to be the fossilized remains of an unknown form of early human life. He was named Eoanthropus Dawsoni after Charles Dawson. For over fifty years the Piltdown man was believed to be the "missing link between apes and man by the majority of scientist. However, later there were questions and controversy over the find and finally in 1953 the Piltdown Man was declared a forgery. 14.

In the October, 2008 issue of the National Geographic magazine, there is a feature article on the Neanderthals that lived in Europe many years ago and became extinct.

This article surprised me because back in the forties it was concluded that the Neanderthals were humans and that they were a race of people that once lived in Europe. They were discovered in France in 1909 and were dated as living 30,000 years ago by the Science Daily, April 15, 2009 issue; while others sources dated them at 300,000 B.C.

There are people living today with much of the same physical features similar to that of the Neanderthals. I knew of one in Chicago and actually met him. He was a professional wrestler and was known as the "Angel". He was very fearful to look at because of his size, but was an interesting person to talk with. A group of scientist carefully studied him and found out, to the disappointment of some, that he was completely human.

In the magazine article, on the subject of Neanderthals, they used some later data for examining the fossils and came to two different conclusions. Some thought the Neanderthals were part human and others thought they were completely human. Evidently, they didn't consider the research done back in the forties. In the magazine, Ed Green, head Biomathematics stated:

> "So the reality is that foremost of the sequence, there's no difference between Neanderthals and modern humans." 15.

Nevertheless, the overall focus of the article was to persuade the reader to view evolution in a favorable way. In the back of the magazine, on page 142, is an article explaining how Adrie and Alfons Kennis created the model of the Neanderthal woman with a spear in her hand and wearing no clothing.

The picture of the woman is on page 36, and since she has a spear in her hand the writer suggested that women must have hunted with the men. They forgot that they were the ones that put the spear in the hands of the woman when they created her.

This is not unusual for those who believe in evolution; to use the image they created as evidence to support their evolutionarily views. For example, in the case of Pithecanthropus Erectus, the lecturer or the writer, at that time, often referred to the large eye brows or the hairy body of the creature, when all they had to go on was the portion of a skull cap, three teeth, and a thigh bone. They used their imagination to create the eye brows and the hair from the drawings and the models they had created!

Getting back to the National Geographic; the article makes the statement that the woman probably didn't wear clothing because the weather was too hot in the summer. Again, this an assumption based upon the model they created with no cloths.

In the study of the science of Ethnology; which is the study of the culture of man; no where, at any time, was it found, that man and woman went around without clothing. The private parts of the body were always covered in every society all the way back to Adam and Eve; and there God made covering for them because they had sinned and were ashamed.

The point here is that if evolution were true; then there would have been a time, when the ape-man would not be wearing clothing. The clothing would come later as the species evolved. Presenting the woman without clothing was an attempt to lead the reader of the magazine, into evolutionary thinking.

The magazine also speaks of the Neanderthal's teeth; that some were worn because the Neanderthals probably used their teeth as a third hand. This is also an assumption. There were probably many Neanderthals with perfectly good teeth and there is no way of knowing that they used their teeth as a third hand. This again, was a suggestion to create the primitiveness of the Neanderthals, to make them a part of evolution.

The article also addresses the large rib cages of the Neanderthals stating it was due to their high levels of activity. But that has little meaning because we have people all around us with large rib cages and it's not because of adaptation, but the results from their parent's genes and the amount of food they eat. Again, the article was slanted and presented in a way to promote the assumptions of evolution. 17.

Next we have Lucy, known as Australopithecus Afarensis. It was discovered in 1973 by Donald Johanson, an American anthropologist, working with the International Afar Research Expedition in Hadar, Ethiopia. They were looking for fossils and artifacts and found the fossil remains of Lucy. They collected 40% of the bones; a fragment from the back of a small skull, part of a femur; some of the vertebrae; part of the pelvis; some ribs and pieces of a jaw.

The sacrum and parts of the shin bone were also found. There were no bones of the hands or feet and it was determined Lucy stood about 3 ft. 8 inches tall and weighed about 65 lb. Many reconstructions have been made of Lucy and they can be seen in different museums in different parts of the world.

Wikipedia, the free encyclopedia has a picture on their web page that shows Lucy's skeleton with just the bones that were originally found. This is the way she is displayed in a Mexican museum.

For the American museums Lucy went through reconstruction. The fragment of the back part of the skull cap and lower jaw bone became a complete ape skull with all its characteristics. In the Cleveland Museum of Natural History Lucy was displayed with constructed hands and feet. The hands were constructed with curved fingers like those of an ape; but they gave her human feet. At the Museum of Natural History in St. Louis they went all out and not only gave Lucy human hands and feet; but also a body with long hair.

The purposes of these changes were for the specific reason to present to the public an image that states the progress of evolution. This exposes the weakness of their theory and the lack of evidence to support it;

> "Though the sacrum was remarkably well preserved, the innominate was distorted, leading to two different reconstructions. The first reconstruction had little iliac flare and virtually no anterior wrap, creating an Ilium that greatly resembled that of an ape."

What they are saying here is that the pelvis of Lucy was at an angle like that of an ape, and it needed to be reconstructed to be more human. As we noted, the picture of the reconstructed skeleton in the Cleveland Museum shows Lucy with ape like hands and fingers; but with human feet. The reconstructed image displayed in the St. Louis museum shows Lucy with both human hands and human feet!

Remember, they did not find any bones of the hands and feet. These reconstructed deceptions should disqualify evolution from ever being considered as a scientific study. 17-19.

In the October, 2009 issue of National Geographic Magazine, it was announced the discovery of the oldest fossil skeleton of a Human ancestor. A research team led by Tim White of the University of California, Berkeley; Berhane Asfaw, former director of the National Museum of Ethiopia; and Giday Wolde Gabriel of the Los Alamos National Laboratory announced the discovery in 1994. The fossils were discovered in Ethiopia's Far desert at the site called Aramis in the Middle Awash region, 46 miles from where Lucy was found

THE CONSCIOUSNESS OF SCIENCE

in 1974. Radiometric dating of two layers of volcanic ash; above and below the find; dated the fossils to be 4.4 million years old.

> "They (the fossils) were so fragile they would turn to dust at the touch. To save the precious fragments, White and colleagues removed the fossils along with their surrounding rock. Then, in a lab in Addis, there searchers carefully tweaked out the bones from the rocky matrix using a needle under a microscope; proceeding millimeter by sub millimeter as the team put it in pieces, the crushed skull and then CT-scanned and digitally fit back together by Gen Tokyo.".

The article does not identify what bones were found. But it does state:

> "The skeleton announced today was discoveredthat same year and excavated with the bones of the other individuals over the next three field seasons. It took 15 years before the research team could fully analyze and publish the skele-ton, because the fossils were in such bad shape."

And later in the article it continues:

> "In the end, the research team recovered 125 pieces of the skeleton, including much of the feet and virtually all of the hands, an extreme rarity among hominid fossils of any age, let alone one so very ancient. The team also found some 6000 animal fossils and other specimens that offer a

picture of the world Ardi inhabited moist woodland very different from the region's current parched landscape."

The same subject is covered in Wikipedia, the free encyclopedia, and states the fossils found were most of the skull, teeth, pelvis, and hands and feet. The fossil was named Ardipithecus ramidus, or Ardi. Great claims are made about this find; that it is more important than that of "Lucy" and that it is the oldest skeleton of a human ancestor.

The final artist's conception and creation looks exactly like a very hairy chimpanzee but with a human body; except for the hands and feet which are complete ape like. Much of the article is about the characteristics of Ardi:

"All previously known-members of our ancestral lineage-walked upright on two legs, like us. But Ardi's feet, pelvis, legs, and hands suggest she was a biped on the ground but a quadruped when moving about in the trees. Her big toe for instance, splays out from her foot like an ape's, the better to grasp tree limbs. Unlike a chimpan- zee foot, however, Ardipithecus's contains a special small bone inside a tendon, passed down from more primitive ancestors, that keeps the divergent toe more rigid. Combined with modification to the other toes, the one would have helped Ardi walk bipedally on the ground, though less efficiently than late hominids like Lucy."

Knowing from the past, how anxious the evolutionists are to find something to support their theory; I question just what it is they found. How accurate are the fossils they found when they had to use a microscope and a needle to "digitally" put the pieces together: "millimeter by sub millimeter"? With 6000 fossils in the area, how certain are we, that what they found, were from the same species.

Notice, they refer to the "primitive ancestors" of Ardi; this is an assumption because they have not found any ancestors; remember Ardi is the oldest.

To show evolutionary trends, we have the statement: "Ardi walk bipedally (two footed) on the ground, though less efficiently than late hominids like Lucy".

How do we know Lucy walked better than Ardi? Remember, they did not find any bones of Lucy's feet, only those that they created for the museums along with the adjustments of the hips to make her more human.

Evolutionist have a history of overstating what they have found and later when the truth is known; they are completely silent about the mistake. They should not be allowed to doctor and camouflage their evidence to support their theory and present it to school children and the public as science! Science, in its true meaning is accepted by Christians with open arms because science magnifies the greatness of our God and His creation. His creation has fascinated man from the beginning of time; science is on our side.

Does Ardi prove evolution? What do you think? I think if evolution were true they should be able to find thousands of Ardis and Lucy's!

In Charles Darwin's book: "Origin of Species", written in 1872; he asks the question:

> "Why is not every geological formation and every stratum full of such intermediate links? Geology assuredly does not reveal any such finely graduated organic chain; and this is the most obvious and serious objection which can be urged against the theory." 21.

What Darwin is asking is why don't we see the evidence of evolution in the fossils found in the stratum of the earth? If evolution were true there should be these intermediate links!

Dr. Colin Patterson who was senior paleontologist of the British Museum of Natural History in his book: "Evolution", did not place any pictures in his book showing any inter connecting species and he was asked why he didn't. He answered:

> "I fully agree with your comments about the lack of direct illustration of evolutionary transitions in my book, If I knew of any, fossil or living, I would certainly have included them...I will lay it on the line, there is not one such fossil for which one could make a watertight argument." 22.

We have to appreciate Dr. Colin Patterson for his truthfulness about evolution: "there is not one such fossil for which one could make a watertight argument."

In Jay Gould's book: "Evolution Now: A Century after Darwin", he wrote:

> "The absence of fossil evidence for intermediary stages between major transitions in organic design, indeed our inability, even in our imagination, to construct functional intermediates in many cases, has a persistent and nagging problem for gradualistic accounts of evolution."

And later in the same book he writes:

> "I regard the failure to find a clear vector of progress' in life's history as the most puzzling fact of the fossil record". 23.

The truth is that the fossil record is not there to support the theory of evolution, Notice in his statements that the evidence is not: "even in our imagination" and the lack of evidence is the: "most puzzling fact of the fossil record" These are not statements of a high school substitute teacher; these are statements of leading scientist in their field that are concerned about their theory.

Jonathan Sarfati, PhD, in his book: "Refuting Evolution", gives many quotes by various scientists who state their questioning the evidence for evolution; but even in their doubts, they still believe in its theory.

For example: Professor D.M.S. Watson, a leading biologist in his day, wrote in his book "Adaptation":

> "Evolution is a theory universally accepted not because it can be logically coherent evidence to be true, but because the only alternative, special creation is clearly incredible." 24.

Professor Richard Lewontin, a geneticist is a renowned champion of neo-Darwinism, wrote in his book: How the World Works:

> "We take the side of science in spite of the patent absurdity of some of its constructs, in spite of its failure to fulfill many of its extravagant promises of earth and life, in spite of the tolerance of the scientific community for unsubstantiated just-so stories, because we have a prior commitment, a commitment to materialism.
>
> It is not that the methods and institutions of science somehow compel us to except a material explanation of the phenomenal

world, but, on the contrary, that we are forced by our prior adherence to material causes to create an apparatus of investigation and a set of concepts that produce material explanations no matter how counter-intuitive, no matter how mystifying to the uninitiated. Moreover, that materialism is an absolute for we cannot allow a Divine Foot in the door." 25.

What the author is saying is: We know we do not have the evidence, and our beliefs at time are a little silly; but we have made our choice and we will stand by it; because it's better than believing there is a God.

Richard Dickerson in his book "Molecular Evolution" wrote:

"Let us see how far and to What extent we can explain behavior of the physical and material universe in terms of purely physical and material causes, without invoking the supernatural." 26.

What he is saying is; with all the evidence we have of the universe, and in our attempt to explain it by evolution; how long, before we must admit there is a God.

FOSSILS and FLOODS

Since millions of years is essential for the evolutionist's theory; we are taught that fossils take millions of years to develop; but we have many fossils that have been fossilized within seconds. A good example that fossilization takes place quickly is in the ruins of the ancient city of Pompeii, Italy. When Mount Vesuvius erupted in AD 79, ash swept though the air and molten lava rushed down the mountain side very quickly, covering everything in its path.

On display at their museum is the fossil of a man drinking from a jug. He was fossilized so quickly the jug was still at his lips when the hot ash covered him. Another fossil is that of two lovers, caught in the embrace of a kiss, frozen in time. These and others are on display at the Naples's Museum. There is also the famous fossil of a large fish caught in the act of swallowing a smaller fish, indicating how swiftly the fish were fossilized.

If fossilization took a long period of time; in many cases the remains of the animal would decay or be consumed by predators before fossilization could take place. Fossils can be found along the shores of lakes, rivers, and oceans, if a person takes the time to look for them. Somewhere in my house, I have a small flat stone that I picked up along the shore of Lake Michigan many years ago, that has the very clear image of a leaf attached to a small stem.

If the leaf had been lying on the open ground, it would have become weather-beaten and wrinkled from the sun. Somehow the leaf was covered by soil, or sand, to retain its shape and in time it became a fossil. The time would depend upon the conditions; but, it would not take millions of years, because the item would not last that long to become fossilized. They have pictures of fossilized jelly fish that had to be done quickly; otherwise, their flesh would not last long on the shore in the sun.

The answer is; fossilization must take place quickly in order to preserve the image. The combination of sand, minerals, chemicals and other ingredients with water, hardens under pressure and becomes rock. This sedimentary rock develops into layers that have developed into mountains. The Grand Canyon is mainly made up of sedimentary rock. When scientists do Geological and Archaeological research at the Grand Canyon they always find the fossils of sea life at all levels of the stratum. This would indicate at one time this area was covered with water and as it receded it left layers of sedimentary rock that developed into the Grand Canyon. You do remember the Biblical record of Noah's flood, don't you?

Some years ago I was working at the Surry Nuclear Power Plant in Surry, Virginia. It is located near the James River and after a heavy rain, shark's teeth can be found along its bank. I learned this one morning when a lady that worked in our office came in wearing a shark's tooth necklace. It was quite a conversation piece.

She told me how she had spent a whole Saturday looking along the bank for a shark's tooth and just as she decided to quit, she bend down to wash her hands in the water and there, right in front of her was this tooth. I asked her how do sharks get this far up the river, because the river is fresh water. Her reply was: "'Oh, the sharks don't come up the river; it's when we have a heavy rain, the water from the mountains washes the teeth down into the river and the current carries them to this area." I asked her; how do the sharks get on top of mountains. She replied, she didn't know! I answered; they got there by Noah's flood!

Archaeologists were digging in the area of Ur, during the years of 1822 to 1834, and came to a level of water laid silk, the slimy green plant life often seen by the edges of lakes. They thought they had come to the beginning of civilization because they no longer found signs of life! There was the temptation to stop; but the decision was made to go on and they dug through eight feet of this compacted water laid silk and below it they found signs of life again.

Dr. C. Leonard Woolley made this statement in his report:

> "it had to take a tremendous amount of water to create eight feet of compacted

water laid silk and the only answer could be "Noah's flood."

This compact water laid silk has also been reported in other parts of the world." 27.

The dating of fossils and rocks is a problem because many times the scientist will want to factor in his evolutionary views in making his conclusions.

Let's take for example that after 500 years we return to the city where we once lived. It has been destroyed and no one is living there. Our purpose is to find out how the people lived at that time. In the excavation many different items are brought to our attention and as we study them, they are labeled with information in regards to the item.

Two of the items have stirred up some questions; one is a very highly polished instrument, and the other is an odd looking tool of some kind that has a long stick with an iron piece at the end with two prongs facing in oppositions directions. It is crude and because of this; the scientist would date it older than the highly polished one. This is because the scientist would look at the two items from an evolutionary point of view; without realizing the pick was used for road work and the polished instrument was used by a dentist.

We have an example of this when on November 15, 2009, and again in December 26, 2010, CBS News: "60 Minutes" presented: The fossils of "Dino Finding, Making Waves"; the finding of "B Rex". Mary Schweitzer, who is a professor at North Carolina State University, told of her experience in testing the bones of "B Rex". She used an acid to remove the outside layer of minerals, and in doing so, discovered the bones had soft elastic tissue. She said:

> "They were transparent Blood vessels. It looks like the soft tissue she would have

> expected to find if it had been modern bone. This is impossible. This bone was 68 million years old. Organic material can't possibly survive even one million years, let alone 68 million."

Notice; there was no thought here that maybe their dating system of "B Rex" might be wrong; its: "How did the tissue last that long!" They assume the dating is correct because it fitted into their theory of evolution of millions of years; therefore, the question was not the age of the dinosaur; but how did the tissue survive? True scientists would be thinking that maybe their dating system might be off.

The fact is that dinosaurs are not as old as the evolutionist would have us believe. This was not the first time that blood tissue has been found in dinosaur bones. 28.

From Answers in Genesis; in Richard Milton book: "Shattering the Myth of Darwinism", he states of a false dating case in South Africa:

> "In one of the most recent cases of anomalous dating, rock paintings found in the South African bush in 1991, were analyzed by Oxford University's radio carbon accelerator unit and were dated to be around 1,200 years old.
>
> "This finding was significant because it meant the paintings would be the first bushman paintings ever found in the open country.
>
> "However, the publicity of the find attracted the attention of Mrs. Joan Ahrens, a Cape Town resident, who recognized the pictures of the paintings as

THE CONSCIOUSNESS OF SCIENCE

being those produced by her art class and were later stolen from her garden by vandals."

"This means the Oxford University radio carbon accelerator was way off on its readings. What is significant about this incident is that in most cases, mistakes in dating cannot be corrected because of the time period; while in this case the actual dates were known!

Therefore in dating millions of years in most cases, we have no time table; no correcting factor to depend upon, or to compare with, to substantiate the date. The date is left entirely to the theories; the assumptions and interpretation of the researchers." 29.

Here is another example of a dating mistake.

"In May 18, 1986, Mt. St. Helens volcano erupted by a 5.1 earthquake beneath the mountain. It removed 1,314 feet off of the top of the mountain and the heat caused the snow to melt and with the molten rock; caused the largest landslide in history as it rushed down the mountain side at speeds of 70 to 150 miles per hour. The debris covered over 23 square miles of area.

The ash cloud reached a height of 80,000 feet in 15 minutes and spread a 300 miles circle from its base. In 15 days the ash had circled the earth. They estimated the temperature reached 660 degrees. The results were that miniature grand canyons were formed in just a very short time and 25 feet of layered sediment covered the area". 30-31

"In dating of the rocks from the Mt. St. Helens eruption it was found that some of the rocks were tested by the potassium-argon method and gave a reading of millions of years; even though the eruption took place in 1986."

In New Zealand, Mt. Ngauruhoe erupted in 1075, 1949, and in 1954; when the rocks were dated; they all came out in the millions of years. How can there be such a difference in the dating?" Dr. Sarfati explains the difference this way:

"What happened was that excess radiogenic argon (40 Argon) from the magma (molten rock) was retained in the rock when it solidified. The secular scientific literature also lists many examples of excess 40 Argon causing dates of million of years in rocks of known historical age. (Like the eruption of Mt. St. Helens, 1986)

This excess appears to have come from the upper mantle, below the earth's crust. This is consistent with a young world; the argon has had too little time to escape. If excess 40 Argon cause exaggerated dates for rocks of known age, then why should we trust the method for rocks of unknown age?" 32.

What he is saying here is, that Argon readings can be faulty when the rock containing it, has prevented its release of Argon because it is contained by the upper mantle of the earth.

Dr. Sarfati gives another example in his book: Refuting Evolution; on pages 111, 112, on this subject. This is where two different dating methods are used and result in two different readings:

> "A piece of wood was burnt from the flow of igneous rock and using radiocarbon dating (14C) came out about 45,000 years old: but when they tested the wood using the K-Ar method the result was 45 million years.
>
> Another example of mistakes in the dating system is the readings of some wood that was buried by lava flow and was charred on one side and dated by radiocarbon (14 C), at about 45,000 years old, but the other side of the wood was dated by the K-Ar method, it was dated at 45 million years old." 33-34.

Without going further; what this all means is that when some new "Ardi" comes along; don't get shook up over it; the truth will come out; God did not create any ape-men. The problem with evolution is not just finding a link between ape and man; it is finding links between each species.

THE SOLAR SYSTEM

On clear nights we are often amazed as we look skyward to see the brightness of the stars and the enchanting light of the moon. Not so much in the city where its light's glare diminishes their brightness; but out in the country where the only light is that from the heavens, and in its quietness.., God almost speaks to us. It is God's handy work and we remember those words: "The heavens declare the glory of God".

The vastness of space cannot be understood or expressed in our normal vocabulary. Even when we see the models of the planets displayed on television or hung from the ceilings of planetariums; we do not grasp the full meaning of how great the space is, that is out there!

Guy Ottewell in his: "thousand-Yard Model" gives us an insight and conception as how great our universe is! He has scaled the solar system down to where we normal people can understand it to some degree! To illustrate the greatness of space he uses these items to present the planets: 35.

1. One Bowling ball to represent the Sun.
2. Three pinheads to represent Mercury, Mars, and Pluto.
3. Two small berries from a pepper plant for Venus and Earth.
4. One chestnut to represent Jupiter.
5. One Hazelnut or acorn to represent Saturn.
6. Two peanuts or coffee beans to represent Neptune and Uranus.

In this object lesson the items are in the scale of 1 inch equaling one hundred thousand miles. This would mean

THE CONSCIOUSNESS OF SCIENCE

one yard (36 inches) would equal 3,600,000 miles. The model shows how overwhelming the evidence is that the Solar System could not have evolved; but had to have an intelligent designer and creator for it to exist.

He explains that the bowling ball is the perfect size for the sun and that it is placed on the floor, or better outside, because plenty of room is needed.

From the bowling ball, measure 10 yards and place a pin in a card, so it may be seen, on the ground and that will represent Mercury. Mercury is the closest planet to the Sun and the distance is 36 million miles. The math would be 10 yards X 3,600,000 = 36 million miles.

Remember, each inch equals one hundred thousand miles; and one yard (36 inches) equals 3,600,000 miles. This is supposed to be simple? Our second stop is 9 yards, so we would measure that distance from the pin in the card representing Mercury and place a new pin with card representing Venus. This would mean Venus is 32, 400,000 miles from Mercury and 67, million miles from the Sun.

From Venus we would mark off 7 yards and place a pepper berry to represent Earth. The distance between Venus and earth is 25,200,000 miles and Earth is 93 million miles from the sun.

Now it gets really interesting because the next planet from earth is Mars and it is 14 yards away. This translates into 50 million miles between Mars and earth and Mars is 142 million miles from the Sun. Mars gets a card with a pin.

Jupiter is the next planet and it is 95 yards from Mars, which translate to 341 million miles. Jupiter is 483 million miles from the sun. It receives a chestnut.

Saturn is 112 yards from Jupiter. Its orbit varies from 941, 070,000 to 840,440,000 miles from the sun so its distant from Jupiter varies. In our scale Saturn would be a Hazelnut.

The next planet from Saturn would be Uranus which is 1,783,500,000 miles from the sun. On the scale it would be 249 yards from Saturn; it is represented by a peanut.

The next planet is Neptune and it is 281 yards from Uranus and it is represented by a peanut. Neptune is 2,793,500,000 miles from the sun.

Pluto is the last of the planets and receives another pinhead. It is 242 yards on the scale from Neptune and 3,671.000,000 miles from the sun. If this model is followed out the total yards would add up to 1,019 yards; or the length of ten foot ball fields.

According to World Book at NASA; "A galaxy is a system of stars, dust, and gas held together by gravity. Our solar system is in a galaxy called the Milky Way. Scientist estimates that there are more than 100 billion galaxies throughout the visible universe. Astronomers have photographed millions of them through telescopes. The most distant galaxies photographed are as far as 10 billion to 13 billion light-years away. A light-year is the distance that light travels in a vacuum in a year—about 5.88 trillion miles.

Galaxies range in diameter from a few thousand to a half-million light-years. Small galaxies have fewer than a billion stars. Large galaxies have more than a trillion. The Milky Way has a diameter of about 100,000 light-years. The solar system lies about 25000 light-years from the center of the galaxy." 36.

Our Solar System covers trillions of miles in space. Each planet is traveling through space on its own orbit, at varies speeds around the sun with exact timing, year after year. There is no confusion; they never run together, they don't bump…; may I ask you a question; could all this be from nothing?

To believe in evolution; you have to believe it all happened! No purpose, no plan, just aimless emptiness with no direction! My friend there is a God in heaven and He wants to know you.

THE SINGLE CELL

As we have said, at one time it was taught that evolution was based upon the development of a single cell. This little creature was known as an Amoeba and could only be seen in a microscope. It somehow came to life, all on its own, and began to divide into a second cell, and then a third, and then on to somehow; it became an organism and through billions of years evolved into the universe as we know it today.

That all changed a few years ago when it was discovered that a single cell is not simple; but, very complex. Today we know that a single cell is filled with all kinds of mechanisms that are complex beyond our imagination.

> "There is enough information capacity in a human single cell to store the Encyclopedia Britannica, all 30 volumes of it, three or four times over." 37.

Jonathan Sarfati, Ph.D.'s in his book: Refuting Evolution, quotes from M. Denton's book; "Theory in Crisis":

> "To grasp the reality of life as it has been revealed by molecular biology, we must magnify a cell a thousand million times until it is twenty kilometers in diameter and resembles a giant airship large enough to cover a great city like London or New York.

> "What we would then see would be an object of unparalleled complexity and adaptive design.

On the surface of the cell we would see millions of openings, like the port holes of a vast space ship, opening and closing to allow a continual stream of materials to flow in and out. If we were to enter one of these openings we would find ourselves in a world of supreme technology and bewildering complexity." 38.

He continues and asks:

"Is it really credible that random processes could be constructed a reality, the smallest element of which a functional protein or gene is complex beyond our own creative capacities, a reality which is the very antithesis of chance, which excels in every sense anything produced by the intelligence of man?" 39.

In Jerry R. Berman's book: The Origin of Life, he explains how the cell, the complex element that makes up all living matter could only exist by creation. The simplest species of bacteria are made up of thousand of genes, complex molecules, DNA, proteins and enzymes, and that the cell could not survive if one of these elements were missing. In other words, one part could not wait around for another part to be evolved; all had to be present instantaneously, at the same time! This could only take place by the process of creation. 40.

Michael Denton is a molecular biologist and questions:

"How could we believe that the cell with its unparalleled complexity could form with out any purpose, without any reason, and without any direction; evolution cannot answer..., with any intelligence at all, demands there is a God!"

DNA

If one has the faith to believe in evolution; he must ask this question: Why would nothing want to produce a cell; and how would nothing do it? Throughout nature fifty percent of the genes come from the male and fifty percent of the genes come from the female. It has always been this way! If the code in the DNA does not have feathers the young will not have feathers. If the parents do not have scales in their DNA the young will not have scales. Each reproduces after its own kind!

The only way a change can take place is through mutation, when this happens the offspring is usually deformed and dies. Mutation is the result of a defected gene that failed in producing the normal offspring. This is known as lost information. It is not the case of a faulty new gene that has been introduced, because that is limited to the genes received from the parents. In other words, no new information can be added to those of the parents; the DNA is fixed at conception from the parents.

Dr. Lee Spetner, who taught information and communication at John Hopkins University in his book: "Not a Chance" wrote:

> "In this chapter I'll bring several examples of evolution, [i.e., instances alleged to be examples of evolution] particularly mutations and show that information is not increased....But in all the reading I've done in the life-science literature, I've

> never found a mutation that added information. All point mutations that have been studied on the molecular level turn out to reduce the genetic information and not to increase it.

> The essential biological difference between human and a bacterium is in the information they contain. All other biological differences follow from that. The human genome has much more information than does the bacterial genome. Information cannot be built up by mutations that loose it. 40

What is being said here is that the human DNA has much more information, capacity to operate than that of an animal DNA. That this information cannot be gained into the DNA of the animal or molecular to allow changes to another species; where would the new DNA come from to start a new species.

(This means the ape to man is impossible because the new genes to make the man can not be added to the genes of the ape. Evolution is impossible!)

The DNA is already set by the parents. Mutation is not new information but the lack on the part of a gene not functioning properly. There is no way for evolution to take place; it's a myth!

Many times the evolutionist presents, what they feel is evidence when it proves nothing. In Newsweek, March 23, 2007, there is a featured article: "The Evolution Revolution" along with a picture of a large skull that would scare and shake the faith of any young believer and maybe a few older ones too.

The article is written by Sharon Begley and is titled; "Beyond Stones and Bones" inferring that: "this new evidence has been discovered that is more convincing in proving evolution than that of stones and bones". What is this new evidence that is so powerful and convincing; ...hold onto your seats...it's "Lice".

> "Head lice live in the hair on the head. But body lice, a larger variety, are misnamed: they live in clothing. Head lice, as a species, go back millions of years, while body lice are of a more recent arrival. (How do they know that?) An anthropologist, Stoneking believed he could calculate when body lice evolved from head lice to body lice by comparing the two varieties DNA, which accumulates changed at a regular rate." 41.

Notice the article starts out by saying: "New evidence has been discovered that is more convincing in proving evolution than stones and bones" and ends up by saying: "comparing the two varieties".

Two varieties of the same species does not prove evolution. If the lice had taken on scales, or feathers, then that would be something! Much of the information about evolution passed off to the public, and especially to the school children, is misinformation designed to sell evolution. (Note: Do they mean now, they have found fossils of lice!)

The variety of a species does not constitute evolution; it simply means there can be many different looking off spring; but they are all of the same species. For evolution to take place; there would have to be added to the species new DNA for wings, or skin, or scales, or shells, or flippers, or feathers; but that DNA would have to be received from the parent's DNA and not by evolving!

This information should be given to young people, who may be struggling with their sexuality; because television and

Hollywood present the homosexual life styles in a humorous and attractive way. The only way homosexuality can grow is by recruitment; they cannot reproduce.

When they make the excuse they were born that way, then the DNA would have to be somewhere in their family background and to say so, would be very upsetting to Grandpa!

In W. Gitt's book: "Dazzling Design in Miniature" he writes:

> "If it's unreasonable to believe that an encyclopedia could have originated without intelligence, then it's just as unreasonable to believe that life could have originated with- out intelligence. Even more amazingly living things have by far the most compact information storage retrieval system known."

> "This stands to reason if a microscopic cell stores as much information as several sets of Encyclopedia Britannica. To illustrate further, the amount of information that could be stored in a pinhead's volume of DNA is staggering. It is the equivalent to the information content of a pile of paperback books 500 times as tall as the distance from earth to the moon, each with a different, yet specific content." 42.

M. Denton's book: "Evolution: A Theory in Crisis", he writes:

> "It would be an illusion to think that what we are aware of at present is any more than a fraction of the full extent of biological design. In practically every field of fundamental biological research ever-increasing levels of design and complexity

is being revealed at an ever-accelerating rate." 43.

What Dr. Denton is saying is that with all the knowledge we have learned about biological design, it is just a fraction of the information that is out there; yet to be uncovered in research.

Evolution is presented in a very powerful way to young minds that are a captive audience from grade school, high school, and college. The plan was well laid out by the Humanist Manifesto II in 1973 and signed by many prominent evolutionists; it reads:

> "I am convinced that the battle for human kind's future must be waged and won in the public school classroom by teachers who correctly perceive their role as the proselytizers of a new faith: a religion of humanity that recognizes and respects the spark of what theologians call divinity in every human being.
>
> "These teachers must embody the same selfless dedication as the most rabid fundamentalist preachers, for they will be ministers of another sort, utilizing a classroom instead of a pulpit to convey humanist values in whatever subject they teach, regardless of the educational level-preschool day care or large state university.
>
> "The classroom must and will become as arena of conflict between the old and the new-the rotting corpse of Christianity, together with all its adjacent evils and misery, and the new faith of humanism...

> It will undoubtedly be a long, arduous, painful struggle replete with much sorrow and many tears, but humanism will emerge triumphant. It must if the family of humankind is to survive." 44.

Here we see the intent and purpose of the humanist, which includes the atheist and evolutionist; that their new faith, their religion, will replace the corpse of Christianity and that the classroom from preschool to the university will be their arena of conflict.

In concluding this chapter I would like to point out the statement Ken Ham makes in his book: "The Lie: Evolution", on page 100, he writes:

> "God made us so that He could relate to us, love us, and pour out His blessings on us, and so that we could love Him in return. On the other hand, if you reject God and replace Him with another belief, that puts chance and random processes in the place of God, there is no basis for right and wrong.
>
> Rules become what ever you want to make them. There are no absolutes-no principles that must be adhered to. People will write their own rules. As the creation foundation is removed, we see the Godly institutions also start to collapse. On the other hand, as the evolution foundation remains firm, the structure built on that foundation-lawlessness, homosexuality, abortion, etc., logically increases."

We have only touched lightly on the evidence that is out there; but all the fields of science nod their heads in approval; that there is a God and He created with wisdom...; all that we see, feel, touch and hear, and He is able to touch and speak to your heart, even right now!

❊ ❊ ❊

CHAPTER III

THE CONSCIOUSNESS OF HISTORY

Does history support evolution? In our last chapter, we gave evidence from various fields of science to show how science supports the Bible, and in this chapter we want to show that history supports the Bible. That it strongly speaks to America's Christian heritage. America was founded upon the Bible and in a Holy God. This Christian foundation did not begin with the landing of the Pilgrims and the investors that came to our shores; but thousands of years ago when God said: "Let there be light!" We start there because that is where the knowledge of God began; in a garden near the Tigris and Euphrates Rivers.

Our nature is such that once our minds are set, it takes a lot of reasoning to change it! It is natural to defend and resist any outside force that would attempt to do that; but to keep abreast of new technique and information, we must be free to weigh the evidence and question the information so that we can make adjustments in our thinking and beliefs. If the evidence does not support our thinking, then it is time to change our belief. Hopefully, the reader will seriously question, not accept, but question the material that's presented.

We have considered the early history of man; that everywhere he has been found, there was the evidence that he had a belief in a supreme being and had an expression of that belief. Just as history records the events and characters in civilization; the Bible provides us with the history of man's relationship with God.

The Old Testament records the creation of the earth and the entire universe of animals, plants, sea life; and yes; dinosaurs too! God created man and woman and placed them in a garden with all the provisions they needed. There was no sickness; no hunger; and no lack of food or entertainment; because the God of heaven met with them and taught them all there was to know about His creation. God said to them:

> "Be fruitful and multiply, and replenish the earth, and subdue (control) it, and have dominion over the fish of the sea and the fowls of the air, and over every living thing that moves upon the earth. And God said, Behold, I have given you every herb bearing seed, which is upon the earth, and every tree, in which is the fruit of a tree yielding seed; you it shall be for meat." Genesis 1: 28-31

THE CONSCIOUSNESS OF HISTORY

The Bible states that God walked and talked with them in the cool of the evening. He gave them a choice; everything in the garden was theirs; but don't eat of that one tree over there! If you do, it will be against my will...and you will die! Of course we know what happened; Eve took the fruit, ate it and gave some to her husband. Adam ate the fruit and as a result sin entered the human race and we see the affect of it every day in our society.

Sin is an act of the heart; it causes us to rebel against God and resist His will for our lives. Since that day, when Adam sinned, we challenge God's will and His right to tell us how we should live! Sin beckons us to experience its excitement; the lights, the music, the laughter, the good times; and then later, we wonder what we did with our lives. Many events that we look forward to with great anticipation, turn out to be boring and when it is over we feel empty. We were made to worship God and we will not find inner peace and satisfaction until we surrender our wills to His.

We have all heard the term: "Born again Christian"; it comes from the verse in the Gospel of John, the third chapter, where Jesus said: "Except a man be born again he cannot see the kingdom of God!" This experience takes place when the person receives Jesus Christ as his Savior and surrenders his will to the will of God: "My life is no longer mine; it belongs to Jesus Christ!"

That is a difficult decision to make; to surrender my plans; my interests; my habits, over to God! Yes, it is! But it's a great trade off, to give up the old life and receive a new life with the inner peace and satisfaction that comes in knowing God. To clarify this, we are not speaking of religion or church; we are speaking of one on one experience with Jesus Christ.

The Old Testament is the record of real people and their lives. They had problems the same as you and I have today! They had families; children to raise, bills to pay, and decisions to make. They lived in different countries with different cultures under kings and rulers, some of them good and some of them bad. Israel was the people God chose; not because they were special; but because: "God loved them, because He loved them".

In reading the Old Testament, we learn that Israel turned away from God many times and they were punished for their disobedience. The surrounding nations, with their idol worship, influenced Israel to intermarry and accept their customs and religion, and in so doing, they turned away from God. In the worship of these other gods Israel was required to offer their own children to be burned alive. We think how horrible it was for them to do such things; but here in America forty five million babies have been killed in the name of free choice because individuals can't, or won't live responsible lives!

There are reports that some of the archaeologists, while translating the writings on the wall of ancient ruins, would become sick to their stomach as they slowly began to understand the meaning of the cuneiforms they were translating. Could what the people did then, be worse than what we are doing today in the abortion clinics?

The little ones are precious to the Lord and His judgments are severe. We covered one of God's judgments on Israel in the chapter on Science, where King Sargon defeated Israel and took them as slaves into Assyria. (II Kings 17: 5-6)

The Old Testament records the genealogy and events of the men and women who lived during that time, and when touring Israel, one can actually, physically, see those places where all the events took place as recorded in the Bible. The Bible is not full of myths!

In spite of all the setbacks by Israel's disobedience, there were those who were faithful and supported God's purpose. When Israel was faithful, God fought for them and protected them; but when they turned away from God, He punished them. The Old Testament contained prophecy concerning the future of those days in which it was written, as well as prophecy concerning the future beyond the days in which we are living. The message of the Old Testament, through those hundreds of years, was that a Savior was coming and He came in the days of the New Testament and its message continues into the days in which we are living now!

In the Old Testament Israel suffered because of her disobedience. In the New Testament the church suffered because of their steadfast faith, love, and determination to serve and live for God.

The history of the Church is that of blood shed and death of innocent people, whose only crime was they loved and lived for Jesus Christ. His message and requirements were simple: separate yourself from all evil; be honest in all your dealings and relationships with other people; and be kind and helpful to those in need.

The Roman Coliseum still stands today, as a memorial and memory of those days, when a crowd of 50,000 people watched and cheered the onslaught of men, women, and children because of their faith in Jesus Christ. It is estimated that in the catacombs, where the Christians lived to escape the horrors of that day, there are the graves of six million persons. The tunnels of the catacombs were on four levels and covered 600 acres. (Wikipedia, the free encyclopedia)

The Emperor of Rome was considered to be a god and all were required to worship him! This, the Christians refused to do, and when they spoke out against the evils of that day, they were persecuted. This persecution of Christians was continued by the Roman Empire into the Inquisitions from 1184 to 1860, which covered 600 years. They are:

> The Medieval Inquisition from 1184-1230
> The Spanish Inquisition from 1478-1834

The Portuguese Inquisition from 1536-1821
The Roman Inquisition from 1542-1860

Christians were considered to be heretics and were persecuted by the priest, bishops and archbishops in the name of the Pope. They were hunted down, tortured, and even killed because they would not recant their faith in Jesus Christ. (This history can be found at Inquisition-Wikipedia, the free encyclopedia.)

In England the Pilgrims were required to worship under the Government controlled church, the Anglican Church of England. They had to accept and worship in accordance to the teaching of the Archbishop and if they didn't, they could be put in jail and even be put to death. They wished to worship after the teaching of the New Testament and not the totalitarian King..; and so it was a ship, the Mayflower, came to America.

They did not come, as some, seeking to conqueror and demand that their faith be the law of the land; they came with the hope, that here in a new land, freedom would be born that came from God and would be known and enjoyed by everyone!

They proclaimed this freedom in their first legal document the "Mayflower Compact"; November 11, Anno Domini, 1620. This document, though short in words, included strong statements of the purpose of their coming to this new land! The Mayflower Compact:

> "In the name of God, Amen. We whose names are underwritten, the loyal subjects of our dread Sovereign Lord King James, by the Grace of God God of Great Britain, France, and Ireland, King, Defender of the Faith, etc. Having undertaken, for the Glory of God and advancement of the Christian Faith and Honor of our King and Country,

a Voyage to plant the First Colony in the Northern Parts of Virginia, do by these presents solemnly and mutually in the presence of God and one another, Covenant and Combine ourselves together into a Civil Body Politic, for our better ordering and preservation and furtherance of the Ends aforesaid; And by virtue hereof to enact, constitute and frame such just and equal Laws, Ordinances, Acts, Constitutions and offices, from time to time, as shall be thought most meet and convenient for the General good of the Colony, unto which we promise all due submission and obedience. Inwitness whereof we have hereunto subscribed our names at Cape Cod, the 11th of November, in the reign of our sovereign Lord King James, of England, France, and Ireland the eighteenth, and of Scotland the fifty-forth.

Anno Domini 1620."

From these statements, we have the purpose of their coming: "to Glorify God and to advance the Christian Faith". We also have the basic fundamental principles and purpose of the government, it is established for that same purpose: "for our better ordering and preservation and furtherance of the ends aforesaid", which refers to: glorifying God and the furtherance of the Christian Faith.

The next sentence refers back to that purpose by saying:

"and by virtue hereof to enact, constitute and frame such just and equal Laws, Ordinances, Acts, Constitute and Offices, from time to time, as shall be thought most meet and convenient for the general good of the colony,..."

Probably the most noted feature Americans can be thankful for; is for the first time a nation was to be formed on the bases of Freedom....; which we are now quickly losing!

In contrast; we might compare the difference between the Pilgrims landing in Massachusetts and the English business men that landed in Jamestown. The Pilgrims; their message and purpose; is still remembered and is alive in America today; while Jamestown, a business investment by the Virginia Company of London is mainly, today, a tourist attraction.

This new beginning that the Pilgrims brought to America was based upon the anointed Word of God, the Bible. Their faith is the same as the early disciples, and those who died in the Coliseum, and those martyred during the Inquisitions. They came and gave us a form of government that the world has never known...and will never know again.

America was to be that land where we could be free; and it was! Our forefathers; the founders of this nation secured that freedom by proclaiming this was to be a country that was founded upon the only source that gave man freedom; the Bible!

Today, this truth is being attacked to the point our citizens do not know America was founded upon the Word of God. As a result of this disinformation from the left; we are seeing the rapid deterioration of our country. If we are honest about this, and I hope "honesty", is the premise for our discussion; we all realize that America is deteriorating. Of course the liberals would say this change is good; that we are maturing as a nation and adjusting to become more acceptable in the world; but this is flawed thinking! An assault is being waged against decency, honesty, morals, respect for the law, and responsible living.

These qualities should be taught to our young people in school; but values can not be taught because then we would be teaching right from wrong; and that would imply religion and would infringe on the separation of church and state. Without moral teaching, young people think it's cool to break the law and not be caught! Rap music and movies encourage the use of dope, (drugs) liquor, sex, and crime. The one shinning light that stands against all this destruction of society is the cross of Jesus Christ; and He is the last person, the world wants to hear about!

In the 60's, America came under the influence of the drug culture. A culture that compromised our basic values of Biblical truths, to values based on a life style with no rights or wrongs. Their programmed thinking said: "No one has the right to tell me what to do." and "if it feels good, do it"! As a result, families were broken up, and lives have been destroyed. "Do as you will!" became the rule of the day. As a result, sons and daughters resent their parents, refuse to listen to their words of wisdom, and turn their backs on the teaching of God's Word.

All our laws, our morals and values came from the Bible. We are not to steal; comes from the Bible. We are not to kill; comes from the Bible. We are not to lie; it comes from the Bible. We are to respect and love one another; comes from the Bible. Just think what a difference the world would be if we just lived by the Bible and the Ten Commandments!

There was a time in America when people didn't need to lock their doors. We lived in Michigan many years ago, and when we went on vacation we did not lock the doors of our home because to do so, would cause the neighbors to think we didn't trust them. This was not uncommon in those days. At that time, I had my own business and delivered my products to the buyer's home. If there was no answer at the door, I simply walked into the kitchen, left the products on the table, picked up the payment, and left. No one would think of doing that today! I think we could agree there has been a big change in our society?

When I was in grade school, our health teacher explained that smoking and drinking were not good for one's health and if we wanted to succeed in sports, we shouldn't do either. The teaching that smoking was bad for us was effective because later in our high school division, only one student smoked and the rest of us thought he was crazy for doing so. Could we deduct from this, that proper teaching about doing the right thing, and the good thing, is beneficial for society? Could we then also deduct from this, that when we do teach what is right and good; young people will not smoke!

Now, we should come to an understanding as to what is true evidence! The information I gave above is true! I know it is true because I lived it; but you the reader do not know it is true; therefore this information is hearsay; not evidence! You may be surprised to learn that much of the information that comes from the left is hearsay; and not evidence. A good example of hearsay is Bob Woodward's book: "State of Denial" where for his documentation he refers to what he heard people say.

We want to make this distinction as to what is truth! Truth must be based upon evidence; not upon assumptions!

A very large myth that is being pushed around today is that America is not a Christian Nation. What they are saying is an expression of what they are wishing; not what the facts prove. The preponderance of evidence that America was founded as a Christian nation makes their statements utterly ridiculous; but still through the media, they are prevailing.

If America was founded as a Christian Nation; what should we find as evidence? There should be written historical documents stating this fact and there should also be visible evidence! Do we have such evidence?

The First Amendment to the Constitution of the United States reads:

> "Congress shall make no law respecting an establishment of religion, or prohibiting the free exercise thereof; or abridging the freedom of speech or of the press, or the right of the people peaceably to assemble, and to petition the government for a redress of grievances."

Although a fifth grade student could understand the above statement; the Judges on our Supreme Court, could not read and understand what the First Amendment meant. It's really quite simple! The first sentence says that the government will not form a religion; and that the government will not prevent the "free exercise" or operation of a religion. Next, that religion shall have full freedom of speech; the use of the press; and the privilege of meeting together! Also, if the government does something to hinder religion; the people have the right to complain about it to the government!

Only twisted thinking could come up with the idea that this meant the separation of the Church from the State. In a nut shell the First Amendment restricts the government from operating its own religion and restricts the government from hindering those who support religion. It is to separate the State from the Church! So the First Amendment gives full freedom for the Church to operate in all of its activities without any interference from the government!

However; what we have today is the government restricting religion, mainly the Christian religion, so it cannot operate in accordance with the freedom of the First Amendment.

It is interesting to note that while our government and nation holds to the separation of Church and state for the Christians; the separation is not applied to the Islam religion. Our public schools can have Muslim Days to celebrate Islam; and even have out side Islam teachers come in for their prayer time; but that would never be allowed for Christians. Also, to meet Muslim needs, we now have footbaths at some of our air ports; and in a city in Michigan, the morning and evening Islamic prayers ring out over loud speakers.

Since this is part of their religion the courts cannot do anything about it; well, prayer and Bible study is part of America's religion too!

The evidence that America was founded as a Christian nation; is in the Colonial Charters signed by Virginia, Massachusetts Bay, Maryland, Connecticut, Carolina, Rhode Island, Pennsylvania, and Georgia. The Charter reads:

> "We, whose names are underwritten, do hereby solemnly, in the presence of Jehovah, incorporate ourselves into a body politic; and as he shall help, will submit our persons, lives, and estates unto our Lord Jesus Christ, the King of Kings and the Lord of Lords, and to all those perfect and absolute laws of his given us in his holy Word of truth, to be judged and guided thereby,"

Here in their signed Colonial Charter are references made to Jesus Christ as the "King of Kings and Lord of Lords" as well as to the Bible as the "Word of truth". One would have to say this charter is very Christian!

Article 22 of the Constitution of Delaware, 1776, required all officers to profess:

> "Faith in God the Father, and in Jesus Christ His only Son and in the Holy Ghost, one God, Blessed forevermore, and do acknowledge the Holy Scripture of the Old and New Testament to be given by divine inspiration

THE CONSCIOUSNESS OF HISTORY

In 1772 Samuel Adams said:

> "The right to freedom being the gift of God Almighty...The right of the colonists as Christians may be best understood by the reading and carefully studying the institutes of the great Law Giver which are to be found clearly written and promulgated in the New Testament."

In 1774, the governor of Boston noted to King George:

> "If you ask an American, who is his master, he will tell you he has none, nor any governor but Jesus Christ." The pre-war Colonial Committees of Correspondence soon made this their motto: "No King but King Jesus."

Remember the criticism John Ashcroft received when he spoke at Bob Jones University and said: "No King but King Jesus?" He was quoting these words of the governor of Boston.

In 1776, the declaration of Independence says:

> "We hold these truths to be self evident, that all men are created equal, that they are endowed by their Creator with certain unalienable rights, that among these are life, liberty and the pursuit of happiness- that to secure these rights, governments are instituted among men..."

In 1777, the First Continental Congress appropriated funds to import for the people, 20,000 Holy Bibles as "the great political textbook of the patriots."

In 1787, George Washington said regarding the Constitution: "Let us raise a standard to which the wise and finest can repair; the event is in the hand of God."

Thomas Jefferson said: "God, who gave us life, gave us liberty. Can the liberties of a nation be secure when we have removed a conviction that the liberties are the gift of God?"

In 1787, at an impass of several weeks at the Constitutional Convention, Benjamin Franklin rose and said:

> "I have lived, Sir, a long time and the longer I live, the more convincing proofs I see of this truth; that God governs in the affairs of man. And if a sparrow cannot fall to the ground without his notice, is it probable that an empire can arise without His aid? We have been assured, Sir, in the Sacred Writings that except the Lord build the house, they labor in vain that build it. I firmly believe this."

He then moved that they resort to prayer.

George Washington always referred to God in his addresses at a thanksgiving Proclamation in 1789 he said:

> "Whereas, it is the duty of all nations to acknowledge the providence of the Almighty God, to obey His will, to be grateful for His benefits and humbly to implore His protection and favor..."

In his Farewell address, George Washington said:

> "And let us with caution indulge the supposition that morality can not be maintained without religion."

THE CONSCIOUSNESS OF HISTORY

Patrick Henry said:

> "It cannot be emphasized too strongly or too often that this great nation was founded not by religionists but by Christians, not on religions but on the Gospel of Jesus Christ."

John Quincy Adams said:

> "The first and almost the only Book deserving of universal attention is the Bible. The highest glory of the American Revolution was this: it connected in one indissoluble bond the principles of civil government with the principles of Christianity."

John Jay, the first chief justice from 1789 to 1795, said:

> "Providence has given to our people the choice of their rulers, and it is the duty as well as the privilege and interest, of a Christian nation to select and prefer Christians for their rulers."

In 1843, Emma Willard, educator and historian said:

> "The government of the United States is acknowledged by the wise and good of other nations, to be the most free, impartial and righteous government of the world; but all agree that for such a government to be sustained many years, peoples of truth and righteousness, taught in the Holy Scriptures, must be practiced. The rulers must govern in the fear of God, and

the people obey the laws...A nation cannot exist without religion. France tried that and failed. We are a Protestant Christian nation, and as such, baptized in blood. Our position ought to be defined as that."

In 1854, The United States House of Congress passed a resolution:

> "The great vital and conservative element in our system is the belief of our people in the sure doctrines and divine truths of the Gospel of Jesus Christ."

Abraham Lincoln said in 1861:

> "It is the duty of all nations, as well as of men, to own their dependence upon the overruling power of God and to recognize the sublime truth announced in the Holy Scriptures and proven by all history, that those nations only are blessed whose God is the Lord."

In 1863, Abraham Lincoln's Gettysburg Address:

> "That we here highly resolve...that this nation under God shall have anew birth of freedom and and that government of the people, by the people, for the people shall not perish from the earth."

In 1892, The Supreme Court of the United States, after citing 87 precedents decided:

> "Our laws and our institutions must necessarily be based upon and embody the

THE CONSCIOUSNESS OF HISTORY

> teachings of the Redeemer of mankind. It is impossible that it should be otherwise: and in this sense and to this extent our civilization and our institutions are emphatically Christian, this is a religious people.
>
> "This is historically true. From the discovery of this continent to the present hour, there is a single voice making this affirmation...we find everywhere a clear recognition of the same truth. These and many other matters which might be noticed, add a volume of official declarations to the mass of organic utterances that this is a Christian nation."

In 1950, President Harry Truman spoke at an Attorney General's Conference and said:

> "The fundamental bases of this nation's laws, was given to Moses on the Mount. The fundamental bases of our Bill of Rights, comes from the teaching we get from Exodus and Saint Matthew, from Isaiah and Saint Paul. I don't think we emphasize that enough these days. If we don't have a proper fundamental moral background, we will finally end up with a totalitarian government which does not believe in rights for anybody except the State".

Woodrow Wilson, in his election campaign for president, said:

> "A nation which does not remember what it was yesterday, does not know what it is today, or what it is trying to do. We

are trying to do a futile thing if we do not know were we came from or what we have been about America was born a Christian nation. America was born to exemplify that devotion to the tenets of righteousness which is derived from the revelations of Holy Scripture."

October 4, 1982, the Joint Resolution of Congress:

"Whereas the Bible, the Word of God, has made a unique contribution in shaping the United States as a distinctive and blessed nation of people. Whereas Biblical teaching inspired concepts of civil government that is contained in our Declaration of Independence and the Constitution of The United States whereas that renewing our knowledge of and faith in God through Holy Scriptures can strengthen us as a nation and a people. Now therefore be it resolved...that the President is authorized and requested to designate 1983 as a national "Year of the Bible" in recognition of both the formative influence the Bible has been for our nation, and our national need to study and apply the teaching of the Holy Scriptures."

I wonder what happened to the separation of church and state back then!

In 1983, President Ronald Reagan also made the above proclamation. President George Bush, in February 3rd, 1983, declared 1990 to be the year of the Bible.

Congressman Randy Forbes, on June 19, 2009, with the approval of twenty-four members of the House of

THE CONSCIOUSNESS OF HISTORY

Representatives presented a Resolution in response to Barack Hussein Obama statement that America is not a Judeo Christian Nation. The resolution also rejects, in the strongest possible terms, any effort to remove, obscure, or purposely omit such history from out Nation's public buildings and educational resources. In his presentation he gave 39 different quotes of statements made by our early founders and political leaders proving America was founded as a Christian nation! These statements are all on public record; they read:

> "All sessions of the United States Supreme Court will start with the Court's Marshal announcing: "God save the United States and this honorable court."

The Federal courts and the United States Supreme Court open with prayer by a minister of the Gospel. The United Supreme Court has declared throughout the course of our nation's history that the United States is a Christian country; a Christian nation; a Christian people: 'we cannot read into the Bill of Rights a philosophy of hostility to religion.'

The most important monuments, buildings and landmarks in Washington, DC, include religious words, symbols, and imagery. The united States Capitol declares: "In God We Trust" prominently in both the United States House and Senate Chambers.

On the top walls in the House Chamber appear images of 23 great lawgivers from across the centuries and Moses, of the Bible, is there and is the only one full face viewed, looking down on the proceedings of the house.

Artwork is found throughout the United States Capitol, including the Rotunda where the prayer service of Christopher Columbus, the Baptism of Pocahontas, and the prayer and Bible study of the Pilgrims are displayed.

In the Cox Corridor of the Capitol are the words. 'America! God shed His grace on thee'; and at the east Senate entrance are the words: 'God has favored our undertakings'

The Ten Commandments are found throughout the Washington area. They are found on the floor of the National Archives; with a statue of Moses in the main Reading room of the Library of Congress; and above the entrance to the U. S. Supreme Court is the statue of Moses with the Ten Commandments in his hands.

The Washington Monument has numerous Bible verses carved in the blocks in the wall with the scripture references: "Holiness to the Lord", "Search the Scriptures" and on the top of the monument are the words: "In God we trust".

The Lincoln Memorial contains Bible verses and the declaration: "This nation under God shall not perish from the earth." In the Library of congress, there is on display the Giant Bible of Mainz and the Gutenberg Bible. There are also Bible verses on the walls of the Library. John 1:5; Proverbs 4: 7 Micah 6: 8; Psalm 19: 1, can be seen and the verse: "The heaven declare the glory of God and the firmament showeth his handy work".

In addition to statements made by the founders of our country and past presidents; every preamble to the constitution of each State makes a reference to God or the Bible. Here are a few of them:

Alaska 1956: "We, the people of Alaska, grateful to God and to those who founded our nation and pioneered this great land."

Delaware 1897: Preamble "Through Divine Goodness all men have, by nature the rights of worshipping and serving their Creator according to the dictates of their consciences."

Illinois 1870: Preamble "We, the people of the State of Illinois, grateful to Almighty God for the civil, political and religious liberty which He hath so long permitted us to enjoy and looking to Him for a blessing on our endeavors."

Iowa 1857, Preamble: "We, the people of the State of Iowa, grateful to the Supreme Being for the blessings hitherto enjoyed, and feeling our dependence on Him for a continuation of these blessings, establish this Constitution."

Maine 1820, Preamble; "We the people of Maine acknowledging with grateful hearts the goodness of the Sovereign Ruler of the Universe in affording us an opportunity...And imploring His aid and direction."

New York 1846, Preamble: "We the people of the State of New York, grateful to Almighty God for our freedom, in order o secure its blessings."

Statements like these are found in the preamble of each state of our Nation. How could our leaders today make such foolish statements as to say we are not a Christian nation? They make these statements based on their own assumptions; and not upon evidence.

All our early universities were founded for the education of our young people to train for Christian service. Out of the first 108 school founded; 106 were Christian. Harvard was set up by its founders believing that: "all knowledge without Christ was vain." Their rules and precepts were: "Let every Student be plainly instructed, and earnestly pressed to consider well, the main end of his life and studies is, to know God and Jesus Christ which is eternal life, John 17: 3 and therefore to lay Christ in the bottomed, as the only foundation of all sound knowledge and learning. And seeing the Lord only giveth wisdom, Let every one seriously set himself by prayer in secret to seek it of Him." Proverbs 2: 3.

"I sought for the greatness and genius of America in her commodious harbors and her ample rivers, and it was not there; in her rich fertile fields and boundless prairies, and it was not there; in her rich mines and her vast world commerce and it was not there.

Not until I went to the churches of America and heard her pulpits aflame with righteousness did I understand the secret of her genius and power. America is great because she is good and if America ever ceases to be good, America will cease to be great."

"In the United States of America the sovereign authority is religious...there is no other country in the world in which the Christian religion retains a greater influence over the souls of men than America."

Alexis De Tocqueville,

No other nation has had the liberty and freedom we have experienced. The difference is America was founded by Christians on the Word of God; the Bible. In closing this chapter I would like to quote from President Ronald Reagan, 1980:

"The time has come to return to God and reassert our trust in Him for the healing of America....our country is in need and ready for a spiritual renewal."

❋ ❋ ❋

CHAPTER IV

THE CONSCIOUSNESS OF THE BIBLE

Many outstanding books have been written by many talented authors; but in time they become out of print and forgotten. Some books with historical records are of value and can still be found in book stores and libraries. The Tale of two cities, for example, by Charles Dickens is one of those books that lives on and has sold over two million copies. However, think of all the books that have been written, down through history and their titles and authors names have been forgotten; and we have no record they ever existed. But, this is not true in the case of the Bible!

The Bible is more popular today than ever before and its history goes back 6000 years. The evolutionist may boast and say their records go back millions of years; but they have no substance to support it; only their assumptions. The Bible has the evidence from science and the testimonies of millions of Christians who have lived in the past and present; that declare the Bible is the Word of God.

In December 18, 2006, The New Yorker published an article titled: The Good Book Business: stating that in 2005, 2.5 million Bibles had been sold in America. That's in one year! This was twice as many books as the current Harry Potter sold at that time. In the list of best selling books by Wikipedia, the Bible is listed as number one with 2.5 billion copies; that's billions, have been sold! The "Little Red Book" by Mao Zedong of China and the "Quran", were tied for second, at 150 million each. The "Lord of the Rings" and the "Book of Moron"; each sold 150 million; and the Jehovah Witness's "The Truth that Leads to Eternal Life" sold 107 Million. The famous communist "Manifesto" has recorded 10 million sold.

According to statistic from Wycliffe International; the Society of Gideon; and the International Bible Society; the number of new Bibles that were sold, given away or otherwise distributed in the United States is about 168,000 per day! The Bible is rightly proclaimed to be: "The World's Best Seller". Perhaps the reader has a Bible or two tucked away somewhere in their home.

A recent pole indicated that 92 % of America homes own at least one Bible and the average household has three; and two-thirds say the Bible holds the answers to the basic questions of life. In 2007, the American Bible Society distributed over 60 million Bibles in the U. S. and the Gideon's distributed 460, 000 Bibles world wide. All of this is a bit staggering...; why do all these people, all over the world want the Bible? What is it about this Book that makes it so demanding? Is it because we do have an inner desire to know God and His message?

In the August 24, 2009 issue of Newsweek; Sharon Begley wrote an article entitled: "(Un) wired for God" in which she writes:

> "There are intriguing neurobiogical findings suggesting that the brain may indeed be wired for God. In additions to the habits of thought that leads us to see the supernatural in the natural and the extraordinary in the ordinary, neuroimaging studies suggest that we came preloaded with the software for belief."

This supports the primus that God has implanted within each one of us the desire to know Him. Do you know this proves there is a God! Evolution could not evolve a desire within the spirit of man to want to know his creator, if there was none! As Sharon Begley suggested; we are wired for God! In contrast we might add; that as much as the Bible is loved and desired; it is equally hated; despised; and attacked!

Today our own government has banned this Book from public schools and government buildings! I have a friend who works on a government project and he is not allowed to read his Bible on his own time, during the lunch hour. For years the Gideon's placed Bibles in the public schools but now, because of the efforts of the ACLU, they cannot do so.

The ACLU, which is government funded, threatens the school board with law suits if they don't honor the First Amendment. Of course our courts have twisted the meaning of the First Amendment and ACLU fail to mention that it states, the government is not to prohibit the free exercise or abridge religion's freedom of speech or of the press, or of their meeting together.

The restriction of the Bible; that is common in communist and Islam countries, has come to the United States.

So how did we get this Book; the Bible, that causes all this controversy by some, and such interest and devotion by others?

The word "Bible" means book, and for the most of its history, it has been known as the "Holy Bible" or Holy Book. To the best of my knowledge, no other author has ever dared to title his book: "Holy". The word: "Holy", carries with it the meaning of Divine reverence and worship; devotion to a sacred God. The Bible was written by forty different people, with various backgrounds, living at different times, spread over thousands of years. This alone is a miracle! Each writer testified that he was not the author; but that he was inspired by God and the Holy Spirit to write what he wrote. The message of the Bible is consistent and in harmony from page to page. It instructs, directs, and foretells coming events.

In the Bible we have the Old and New Testaments. The word, "Testament" means covenant. The Bible is then a covenant of God made with man at the very beginning of time. From the Bible we have our twenty four hour days; our seven day weeks; and our twelve months to the year. Our years are dated from the birth of Jesus; and our present day laws, our morals and instructions on how to live, also come from the Bible! Living by the Bible gives us inner peace, contentment, harmony, and success; living outside the Bible, gives us crime, distress, wars, and confusion. From this we can deduct America is certainly not living by the instructions of the Bible today!.

The writings of the Old Testament were kept and preserved by the Jewish priests. They were written on parchment, made into scrolls and were considered sacred documents that were worshiped by the Jewish Nation. The scrolls were the inspired Words of God.

The New Testament was written by the disciples and followers of Jesus Christ. They were with Jesus for three years as He ministered to the people. They talked with Him, and ate with Him as they traveled through out the regions. They saw the miracles He did; the feeding of the five thousand; the healing of the sick; the restoring sight to the blind; the raising Lazarus from the dead; and the calming of the storm when they were in that boat on the sea of Galilee.

They listened to His teaching; and when the time came for Him to be crucified; they were there and witnessed His cruel death. On that first Easter morning, they were there when He appeared and showed them His wounds to prove He was alive. He remained on earth for forty days after His resurrection and was seen by five thousand people before He ascended into heaven. When the time came to write the New Testament; the evidence used was overwhelming, because there was so many who testified to the truth of the accounts that took place. Written documents of letters, statements of the disciples, and interviews with the people, compiled the information that would be considered. Much heart searching and prayer went into the process; because what the early church was doing was creating God's Word.

The early Church prayerfully studied each piece of evidence to determine what was consistent with the Old Testament and the teaching of Jesus Christ. When the testing and studying was finished; the church fathers accepted the information and because it met the strict requirements and evidence it was canonized in 100 A.D. as the inspired Word of God; The New Testament.

We have over 4000 known manuscripts of the Bible. That is more evidence than we have of any other author; or book, past or present! The Bible's accuracy is based upon ancient manuscripts, written copies and translations.

The ancient copies from original manuscripts include the Codex Sinaiticus; the Codex Alexandrinus; the Samaritian Pentateuc; the Peshito of Sytiac; and the Vulgate. All of these date back within the first 400 years A.D. Our modern day translations are created from the careful study of these documents along with the language the documents were written in, to preserve the text, and present to the reader a Bible that will convey its message in a clear, understanding and meaningful way.

The Bible takes us back before time began; before there was a heaven and earth as we know it. There was an argument between the Godhead and some of the angels. Lucifer challenged the authority of God and God responded back; "I can create something out of nothing that can defeat you!", and Satan answered: "You're on!" Did God know at that time, that Adam would sin? That His creation would turn against Him and cause Him grief? Yes, He did! Then why would He go through with all of this?

God knew that there would be a group of people that would receive His son as their Savior and out of their own free will, would love and serve Him. We, who know Christ, worship Him not because of the benefits we receive, or the joy of living for Him each day, because even if there were no heaven to look forward to, and death ended it all, we would still love Him; because we came to know Him! In 1st Peter 1: 8 we read:

> "Whom having not seen, we love; in whom, though we see Him not, yet believing, we rejoice with joy unspeakable and full of glory".

Before creation started and earth's foundations were laid; God knew there was a cross. When Adam sinned; the innocents of his relationship with God was broken. Because of this, with the inner desire to know God, came the inner desire to turn away and to rebel against God. The Biblical term for this is referred to as the Old Nature and the New Nature; or the Old Man and the New Man.

When we accept Christ as our Savior, the Spirit of Christ comes into our heart and changes our old nature that rebelled against God; to the new nature that loves God. You remember the verse I gave before: "If any man be in Christ He is a new creation..." II Corinthians 5: 17. Christianity is the only religion that offers a new beginning in life; the slate is wiped clean and one can start all over..., with God! This is what Brit Hume was trying to say to Tiger Woods; that Christ offers a new beginning that no other religion can offer.

The Cross, that is taken so lightly today, was the implement God used to defeat Satan and redeem His creation. Mel Gibson's film: "The Passion of the Christ" brought realism to our minds the horrible torture and pain that was inflicted upon Christ, when He died upon that cross. But the film did not portray all that took place on that cross or in heaven. For thousands of years the Jewish people sacrificed lambs and offered its shed blood on the altar to God to receive forgiveness of their sins. This they did every week.

In Christ was the perfect sacrifice because He was sinless and He offered His blood, as the Lamb of God, on the altar of the cross, to pay for our sins; so we can have forgiveness. The death of Christ did more than just pay for our sins, because when we accept Christ as the one who died for us, the record is deleted and replaced with a clean white sheet. We are declared righteous! This is the New Covenant; the New Testament stained with the blood of Jesus Christ!

Blood is not a pleasant subject to talk about; but it was the only way God could judge, punish, and condemn sin, once for all..; and make it possible for us to be acceptable to Him. He did it with Christ's death on the cross.

In II Corinthians 5: 19, 20; we read:

> "To wit that God was in Christ reconciling the world unto himself, not imputing their trespasses unto them; and hath committed unto us the word of reconciliation, Now then we are ambassadors for Christ, as though God did beseech you by us, we pray you in Christ's stead, be ye reconciled to God. For He hath made Him to be sin for us, who knew no sin that we might be made the righteousness of God in Him."

In this verse notice that God is not holding the sinner responsible: "not imputing their trespass on them", He was placing their trespass, their sins, on Christ. Notice it also states that: "Christ who knew no sin was made sin for us...that we might be made the righteousness of God in Him!" Did you understand that? Christ was made sin! That through His shed blood and its cleaning power we are declared righteous when we accept Christ as our Savior. We make the decision to accept or reject what Christ has done.

Let me give an example in real-estate as to what Christ did for us. When a couple decides to buy a new home and have found the one they desire to purchase, they sign a contract with intent to buy and it is presented to the seller for acceptance. If the seller agrees, he signs the contract and then an arrangement is made for a closing date on the property. This is usually done in a lawyer's office.

At that time, the seller and the buyers will meet with the lawyer and when all agreements are met, there will be an exchange of money from the buyers to the seller and the deed of the property will be conveyed, imputed, from the seller to the buyers. The buyers now become the owners of the property.

THE CONSCIOUSNESS OF THE BIBLE

This is what took place when Christ died upon the cross. Our sins were placed upon Jesus, they were imputed, conveyed to Him and His righteousness was imputed, conveyed to us. In Jesus Christ, we become the owners of His righteousness....and He owned our sinfulness. In this we have complete forgiveness and the gift of eternal life.

If you have never accepted Christ as your Savior, it is simple. Don't need a priest or a minister; its one on one. Just tell God you are now accepting Jesus Christ as your Savior and thank Him for dying for you.

In the days of the early church, copies of the Bible were hand written so few members had the privilege of owning one. The New Testament message depended upon the preaching of the Word and most of that was done from the Septuagint Bible.

The language was changing from Greek to Aramaic and then to Latin. By 400 AD several versions of the Bible had been made; one of which was the Vulgate that was very popular with Christians and used mainly until the fifteenth century. It later became the Bible that the Catholic Church used for many years.

John Wycliffe was a Catholic priest who decided the English people should have the Bible in their own language and against the priest's wishes, translated the Latin Vulgate into English. The Catholic Church felt the people would not be able to understand the scriptures and wanted to stop Wycliffe from translating the Bible. Wycliffe completed his translation and it became so popular the priest were afraid to take action against him; so after his death they dug up his body; burned it and threw the remains in the Swift River.

William Tyndale was also a Catholic priest who made a translation from Latin into English, as well as German. He also was in trouble with the church because they did not want the Bible translated into the German language of the people. Near the end of his life he was arrested and charged as a heretic and when he would not recant he was hanged and his body was burned at the stake.

In 1535 Miles Coverdale, an English Bishop completed a version of the whole Bible that was called the Coverdale Bible. He was commissioned by the Archbishop of Canterbury to publish the Great Bible which was fourteen inches thick. The king ordered the Great Bible to be place in each church and this started the opportunity for the common people to read the Bible.

The next translation was the Geneva Bible in 1560 which was the first to divide the Bible into chapters and verses. It was also the Bible the pilgrims used.

The Bishop's Bible was printed in 1560 under the direction of the Archbishop of Canterbury during the reign of Queen Elizabeth.

The Douay Bible was a version of the Latin Vulgate published especially for the Catholic Church and completed in 1582.

In 1611 the King James Version of the Bible was printed and it is probably the most used and loved translation of them all. It was authorized by King James I of England and was translated by forty-seven scholars using the Hebrew and Greek texts along with other translations and the Bishop's Bible. It projects the message of the Bible with a power that has been lost in the later paraphrased translations.

Following the King James Version is the Revised Version, 1881-1884; and The American Standard Version, 1900-1901.

In 1964 the Amplified Bible was printed; it differs from other translations in that the shads of meaning from the original languages are used to bring out the full meaning of the text.

THE CONSCIOUSNESS OF THE BIBLE

The Living Bible was printed in 1996; a paraphrased translation putting the Bible into modern day language, making it easier to understand. Other paraphrased translations have followed, such as the Message in 2002 that simplifies the language of the Bible to every day conversation.

One of the problems with the paraphrased translations is the inability for the Christian to quote the sentences and memorize them. We can say their meaning; but we can't quote them because they are in sentences instead of statements. Knowing what God said and hiding it in our hearts is a very important factor in living our Christian life. "Thy Word have I hid in my heart that I might not sin against thee." Psalms 119: 11. There are times when the Christian must call upon the Word of God for reassurance and strength to meet a problem in his life; and it's the knowledge of God's Word that gives us the victory under those circumstances. His Word has power; and we must have the Word of God in our hearts.

The Bibles own explanation as to how it came about is found in II Peter 1: 21:

> "For the prophecy came not in old time by
> the will of men; but holy men of God spoke
> as they were moved by the Holy Spirit."

The Bible encompasses the complete history of God's love for His creation and ends with an invitation in Revelation 22: 17:

> "And the Spirit and the bride say come.
> And let him that hears say come. And let
> him that is athirst come. And whosoever
> will, let him take of the water of life freely."

This Book.., this Book, that has been burned, banned, smuggled at the risk of imprisonment and death in many countries, can be bought for a dollar in a Dollar Store here in America! Do you see that book over there? Maybe it's on a shelf, or on a table; perhaps somewhere in another room; go get it and give it a hug! It is the Word of the living God.

❈ ❈ ❈

CHAPTER V

THE CONSCIOUSNESS OF THE HOLY SPIRIT

You may ask what has this subject to do with evolution being a myth. It has to do with truth! The students taking a class on Biblical Studies at Rice University under Dr. April DeConick will learn:

> "Jesus is nothing more than a constructed person who exists only in our imagination. What bothers me is I don't think Jesus is interwoven into the history of the early first century. We know that Jesus is a type

of figure people in that time and culture made up, so in the absence of historical evidence for Jesus, I think we should be agnostic, if not slightly suspicious of Jesus existence."

(From Dr. DeConick's web page)

Such a statement is difficult to understand with all the evidence we have in regards to the reality of the person of Jesus Christ. To say there is an "absence of historical evidence of Jesus," is to question history itself! Dr. DeConick seems to be teaching myths.

For six thousand years, multimillions of men and women have committed their lives to serve the living God; willing even to die horrible deaths rather than recant their faith. By contrast, in the last two hundred years society has accepted a tooth, a portion of a skull cap, three teeth, a leg bone, the remains of some tiny creature put together with a lot of imagination; to cancel out the Bible and the belief in God. Which carries the most evidence? Which would you believe?

The truth of the matter is that the teaching of evolution has entered into our Christian colleges, high schools and churches. The twenty-four hour days of the Biblical creation is supported by science; periods of time would be disastrous for plants and animals to survive. The Bible is clear about the meaning of "Day" because in Exodus 20: 10, 11, it states that God created in six days and rested on the seventh day. This is scripture supporting scripture.

Throughout history the seventh day of the week has been known as the Sabbath Day, the day set aside by God for the Jewish people to worship Him. That day is still honored today, as the Sabbath and if you look on the calendar you will see it is still the seventh day.

The person behind the pulpit and in front of the class room has the responsibility to present truth and to teach otherwise is to deceive those who are listening!

Evolution is based upon assumptions, a belief system that is still seeking some kind of solid evidence to support it. We have presented the statements from scientists from different fields that have stated in their own words, concern about the lack of evidence to support their theory. The Bible on the other hand remains the most popular book in the world outselling all others. There has been no scientific proof to discredit it, it gives eye witness accounts of the history of the world, and its message has changed the lives of millions of people in the past and present. With all assurance the Christian can declare: "I know that my redeemer lives!"

Although, this subject is more directed to Christians; it may be of interest to others, because it deals with the inner power and working of the Holy Spirit, the third member of the Godhead family. By introduction, let me say that this is a subject that is the least understood and most avoided by ministers and many Christian schools. Not because the Bible is not clear in its presentation of the subject; but because we are not willing to accept what the Bible says on this subject!

The failure of the Bible believing minister to proclaim the complete Word of God has led to compromising the standards by which we live. Fornication is no longer considered to be a sin and so living together is well accepted while the holy institution of marriage is rejected. Relationships have become so fickle that now agreements are written up as to who will get the cat or dog in case the couple decide to break up!

Much of the problems we face in America; the high crime rate, the decline in morals, sexual freedom, and disrespect for the law, lack of direction and responsibility, and you name it; can be partly laid at the feet of American pulpits and congregations.

We have been content to sit each Sunday in church and play out our role of being Christians while the world is systematically destroying our influence and Biblical truth. What is wrong with us when the public think Christianity is on the same level as other religions? Have we become politically correct or are we just compromising our message?

Bit Hume was verbally attacked because he recommended Christianity to Tiger Woods to help him recover his life. He was accused of presenting Christianity as being better than Buddhism.

It was said of Jesus, that when He came; "He would baptize with water and with fire!" When Alexis De Tocqueville sought to find out what it was that made America so great; he said it was "the pulpits aflame with righteousness." Jesus said: "you shall receive power, after that the Holy Ghost shall come upon you and you shall be witness unto me".

It is the power of the Holy Spirit living within us that enables us to live the Christian life. "I believe in Jesus"..., will not do it! The devil does that! Our decision in becoming a Christian must come with the understanding and the appreciation; that without Christ, our sins will condemn us to hell!

Playing church on Sunday is not going to make it! We may have our name on a card that makes us a member of a church; but it will not get us into heaven! Only the shed blood of Jesus Christ can get us there! Jesus said in John 3: 3 "Except a man be born again, He cannot see the kingdom of heaven."

Maybe these are strong words for the reader, and they are strong words; but we are living a life here, and God wants us to live it in health, freedom, and with the blessings that come from knowing and living for Him. The only way to know God is by knowing Jesus Christ! Jesus Said: "I am the way, the truth, and the life: no man comes to the Father, but by me", and He meant that! Life is not just a game to be played; because someday it will all come to an end.

On a grave stone in a small cemetery in Palatka, Florida, there are these words of a mother speaking to the grave of her nineteen year old son, saying: "Good by my friend, this is the end; I will never look into your eyes again.", and then in poetry form on the other side are the words: "good-by my son, into the black midnight". I have been to Palatka several times and I cannot leave that city without visiting that grave that expresses such hopelessness.

Robert Ingersoll, a noted atheist at his brother's funeral, could not have a minster attend because he didn't believe in God; and so he stood by the grave and said:

> "Whether in mid-sea or among the breakers of the farther shore, a wreck must mark at last the end of each and all. And every life, no matter if its every hour is rich with love and every moment jeweled with joy, will to its close, become a tragedy as sad, and as deep and dark as can be woven of the woof of mystery and death."

Led Forth in Triumph", by D. James Kennedy, Ph.D.

I don't remember who said it; but it's worth repeating:

> "We are not human beings going through a temporary spiritual experience; we are spiritual beings going through a temporary human experience!

Merrill Womach, a Christian man who taught music and sang at concerts all over the world, was taking off in his airplane when it crashed and burst into flames. He was badly burnt all over his body. In the ambulance on the way to the emergency room he regained consciousness and instead of screaming in pain, as the attendees expected.., in his beautiful tenor voice he sang:

> Wonderful peace...... Wonderful peace.
> Peace that the world cannot give.
> When I think how He brought me from darkness to light,
> There's a wonderful...., wonderful peace.

He lost his nose, his lips, his eyebrows, his eyelids and his head was swollen to the size of a basket ball. We met him some years ago when he sang at a yearly meeting in Fulton, Missouri. He was a shock to look at; but the message he brought in word and music will never be forgotten. He joked about his physical appearance saying: "I never have to dress up for Halloween."

This inner assurance and strength comes when our heart and life has been changed by the power of the Holy Spirit dwelling within us through a relationship with Jesus Christ. We are not religious imitators! Christ did not suffer the agony of the cross so we could play religion! He died there because He loved us and wanted us to experience all the blessing of heaven here on earth. The surrendering to Christ is a hard and difficult decision to make until we realize what He did for us upon the cross.

Jesus had told His disciples that when He went away He would send the Comforter, the Holy Spirit. He would be a helper, counselor, advocate, intercessor, and strengthener in their lives. That He would take the character, attitudes, and personality of Jesus Christ and create them in us.

THE CONSCIOUSNESS OF THE HOLY SPIRIT

That: "we would know Him, for He will dwell with us and would be in us." John 14: 16, 17

After the resurrection of Jesus Christ from the dead and His ascension into heaven, His disciples waited in Jerusalem, as they were instructed, to receive power from the Holy Ghost. The account is recorded in the Book of Acts, chapter 2: 1-4:

> "And when the day of Pentecost was fully come, they were all with one accord in one place. And suddenly there came a sound from heaven as of a rushing mighty wind, and it filled the entire house where they were sitting. (3) And there appeared unto them cloven tongues like as of fire, and it sat upon each of them. (4) And they were all filled with the Holy Ghost, and began to speak with other tongues, as the Spirit gave them utterance."

(The word: Ghost, is the old English word for Spirit)

Verse 4 states that they were all filled with the Holy Spirit, and began to speak with other tongues as the Spirit gave them utterance. When this took place, I can't help but believe there was an emotional overflow of great praise and rejoicing among the Christians. This was a confirmation of the reality of their relationship with God and Jesus Christ. Romans 8: 16: "The Spirit beareth witness with our spirit, that we are the children of God"

The word tongues could be translated languages because in many cases they are the languages of other nationalities; but also they are special communications with God through the Holy Spirit. We are told in the 14th chapter of 1st Corinthians that when we pray in the Spirit, God understands us and that in so doing we are built up spiritually in our faith.

Notice that in verse 4, those who spoke, did so as the Spirit gave them utterance. This means they were not all speaking at once; but there was order as to who was speaking and what was being said. The Holy Spirit was overseeing the meeting. Their praise and worship in different languages would certainly get the attention of the people.

Pentecost was a special celebration that brought the Jewish people to Jerusalem from many foreign countries. When the members of the early church became aware of the crowd they directed their conversation to the people. We know this because in verses 7 and 8 in the same chapter, it states that when one of the disciples addressed the people, he spoke in the language of the people from the country of that language: (verse 7)

> "And they were all amazed and marveled, saying one to another, behold, are not all these which speak Galileans? And how hear we every man in our own tongue, wherein we were born."

So then, the disciple or person speaking, even though he didn't know the language of a certain group of people, spoke their language. The disciples were bringing a message from God to the people. We learn in verse 11 that several of the disciples or apostles spoke, because in naming the different countries the people were from, it states: "We do hear "them" speak!" From this verse we also know the subject the disciples were speaking about; they were speaking about: "The wonderful works of God".

If someone was standing there and heard the person speaking in a language he did not understand, or even more, in a language he never heard before, he might think they had too much to drink. I have listened to Koreans speak and have said to myself, how can they possibly be communicating. Peter spoke and when he finished, the people were so moved that three thousand of them accepted Jesus Christ as their Savior! Acts 2: 14- 36:

> "When they heard this the people were pricked, stung, in their hearts, and said to Peter and the rest of the apostles, Men and brethren, what shall we do? Then Peter said unto them, repent, and be baptized every one of you in the name of Jesus Christ for the remission of sins, and you shall receive the gift of the Holy Ghost. For the promise is unto you, and to your children, and to all that are afar off, even as many as the Lord our God shall call."

Verse 41 states:

> "Then they that gladly received his word were baptized: and the same day there was added to the church three thousand souls."

This was the beginning of the operation of the Gifts of the Holy Spirit in the lives of individual believers. The Holy Spirit was very much at work during the days of the Old Testament; but now He was to begin with His special Gifts in the Church.

Many of the denominational churches have taken the position that the Gifts were given just to help the growth of the early church while the New Testament was being written; and after that, the Gifts were to be discontinued. What evidence from the Bible do we have that the Gifts really stopped?

Let's look again at the verses in Acts the second chapter. In Peter's message he reminds the people of their history going back to David and the promises of the coming of

Jesus, whom the people crucified. In Acts 2: 38, Peter proclaims that if the people will repent from their sins, they would receive the Gift of the Holy Spirit. Then in verse 39 it states:

> "For the promise is unto you, (the promise of the Gifts of Holy Spirit) and to your children, and to all that are afar off, even as many as the Lord our God shall call,"

This verse states that the promise of the Gifts of the Holy Spirit is for "you and your children". The New Testament was completed and canonized by 100 A.D. The children, the next generation that Peter spoke of, would live beyond that date when the New Testament was canonized; therefore, the Gifts of the Holy Spirit would extend beyond the day the New Testament was completed! There is no scriptural evidence to support that the Gifts were to stop when the New Testament was completed. Some where, some how, false teaching and preaching made its way into the Bible believing church, and because of lack of faith in the Word of God, human reasoning was added to the scriptures.

This is especially true today where our Christian schools and Christian leaders have compromised the Word of God to allow the teaching of evolution to be applied to the days of creation. It sounds so much better on national television to say: "I don't know if the days in Genesis were twenty four hours days or periods of time, I wasn't there!" This way you keep the atheist happy, and come off sounding like an open minded person which is more acceptable to the public!

There is this fear about receiving the baptism of the Holy Spirit.., that is what we are talking about, isn't it? We have this fear we may lose control; we have heard and seen movies of the ugly display of emotions that people have in religion and we don't want any part of it! Let me say something here; the Holy Spirit never takes over a person and makes them out of control in an unseemly manor.

I have known the Gifts of the Spirit in my life for over forty years and only once did I ever see a person get all excited and go running wildly around on the stage, and that was because this person, who hadn't been able to walk in ten years, was miraculously healed. What will it be like for those ministers and teachers, who have preached and taught against the Gifts of the Holy Spirit throughout their lives, when they stand before God?

In addition, the verse states that the promise of the Gifts of the Holy Spirit is for those "afar off, even as many as the Lord our God shall call." God is still calling today; and so the Gifts are still for us today! If the reader is a Christian, then you know the Word of God is to come first in your life!

The objection of the baptism of the Holy Spirit by the denominational preachers is that we receive the Holy Spirit when we are saved, which I believe is true; but there are Biblical accounts where the believers had accepted Christ as their Savior, were water baptized, and then receive the Holy Spirit later. The Baptism of the Holy Spirit equips us for service.

In the Book of Acts the 8th chapter is the account of the church in Jerusalem being persecuted and the church fleeing the area for safety because Steven had just been stoned to death by Saul.

In the 5th verse of Acts 8; Philip goes down to Samaria and preaches to the people. The message and miracles were so strong that the people responded and the whole city turned to Christ. We know there was a great response because the scripture states there was great joy in the city. (The Bible says: the joy of the Lord is our strength!)

In the 12th verse it tells us that the people believed in Jesus Christ and were all baptized; job completed? No, because when the apostles in Jerusalem heard that the people in Samaria had received the Word, Peter and John went down and prayed for them to receive the Holy Spirit. Remember these believers had already accepted Christ and had been water baptized. What happened to: "we receive the Holy Spirit when we receive Christ?"

In Acts the 19th chapter Paul visits Ephesus and finds disciples there and in fellowshipping with them, feels there is something wrong and asks them: "Have you received the Holy Spirit since you believed?"

I have heard the argument that these were not really Christians; but if they were not Christians I believe Paul would be seeking that they first come to Christ; but their salvation was not in question. Paul's concern was their experience with the Holy Spirit. "Have you received the Holy Spirit since you believed?" So they were already believers! They were Christians, but had not received the baptism of the Holy Spirit, so Paul laid his hands on them and prayed and they received the Holy Spirit and began to speak and pray in the Spirit; tongues if you please.

This is another case where there were believers, Christians, and later received the baptism of the Holy Spirit. It is this baptism that units the body of Christ. In I Corinthians 12: 13 it states: "For by one Spirit are we all baptized into one body, whether we be Jews or Gentiles, whether we are bond or free; and have been all made to drink into one Spirit." Water baptism does not unit us; but the oneness of the Spirit does. My background is Baptist but some of my most precious fellowship has been with the Spirit filled Catholics.

We have more evidence that the Gifts of the Holy Spirit are for today because in 1st Corinthians the first chapter, verses 4-7. Paul is writing to the Corinthians and states:

> "I thank my God always on your behalf, for the grace of God which is given you by Jesus Christ. That in every thing you are enriched by Him, in all utterance, and in all knowledge; Even as the testimony of Christ was confirmed in you; So that you are not lacking in any gift; waiting for the coming of our Lord Jesus Christ."

Here Paul is saying that the Corinthian church has everything they needed, while they are waiting for the return of Christ. They are not lacking in any gift, and if we look at the original meaning the translation would read: "lacking in any spiritual endowment, due to the power of divine grace operating in your souls by the Holy Spirit". These Gifts, the church was to experience while they were: "waiting for the coming of our Lord Jesus Christ". Since Christ has not yet returned, then the Gifts of the Spirit must still be for today.

There has been criticism, by some, that Paul is rebuking the Corinthian church in the 3rd chapter of 1st Corinthians because they were abusing these gifts and referred to them as being carnal and babes. He called them carnal and babes, not because they were speaking in tongues; but because in the household of Chloe, there was dissension and divisions among them; some supported Apollos and some supported Paul. 1st Corinthians 1: 10-12

In Luke 11: 1-13 Jesus is speaking to His disciples in answer to their request for Him to teach them how to pray. He then gives them what is known as the Lord's Prayer. Afterwards, He explains God's willingness to answer our prayers; that we should be insistent and consistent in our prayers; that we are to keep knocking! He taught that God will not give us something bad; like a father would not give a stone to his child instead of the bread that the child needed. And then in verse 13 it states:

> "If you then being evil, know how to give good gifts unto your children: how much more shall your heavenly Father give the Holy Spirit to them that ask Him?"

Here we have the Lord's Prayer, instructions on how to pray, and how to receive the Holy Spirit in the same chapter of the Bible. This chapter is not presenting a choice; that prayer and instructions are ok, but the Holy Spirit is not! They are presented together for us to know, to understand, and apply to our lives. If the Lord's Prayer is still for us today, why aren't the gifts of the Holy Spirit for us today?

The non believers in the Gifts of the Holy Spirit, may say, well, the Holy Spirit, we need in our every day living, as part of our Christian experience...; then why not His Gifts also? This verse also brings up another question. We know we receive the Holy Spirit when we receive Jesus Christ as our Savior, so why then, in this verse must we ask for the Holy Spirit?

The truth that the Gifts of the Holy Spirit are for today is clear in the 13th chapter of 1st Corinthians. If you read the ending of chapter 12 you will see chapter 13 is the continuation of chapter12. The verses would read this way: "But covet earnestly the best gifts: and yet I will show you a better way. Though I speak with the tongues of men and of angels, (the gift of tongues) and have not love, I am become as sounding brass or a tinkling cymbal. And though I have the gift of prophecy, (another gift of the Holy Spirit) and I understand all mysteries, and have all knowledge; (the gift of knowledge) and though I have all faith, (the gift of faith) so I could remove mountains, and have not love, I am nothing."

What Paul is saying here is that if we are using the Gifts of the Holy Spirit for any reason outside the motivation of love, it will be of no profit or gain. That even feeding the poor or other social work, using our time and energy for good without doing it in love; means nothing to God. That love conducts itself in an admirable way; love endures suffering, is kind, is not envious, does not glorify self, is not easy provoked, and thinks no evil. The Gifts are to be administered in love. If we do not minister the Gifts in love; then our ministering will be as the sound of brass or tinkling cymbals.

There are those who say that the 13th chapter of 1st Corinthians teaches that the Gifts of the Holy Spirit stopped when the New Testament was completed. Ok, let's look at those verses. One of the verses is 1st Corinthians 13: 8:

> "Love never fails; but whether there be prophecies, they shall fail; whether there are tongues, they shall cease; whether there be knowledge, it shall vanish away."

The verse is clear! There is a time coming when these three gifts: prophecies, tongues, and knowledge will cease, as well as the other Gifts; when we are in heaven. In heaven all prophecies will be completed; tongues, praying in the Spirit will be unnecessary; and knowledge, we will know Him as He knows us. The critic's view is that these Gifts stopped at the completion of the New Testament; but it is obvious they didn't because knowledge didn't stop and since it has continued so must the other Gifts.

Then we come to verses 9 and 10:

> "For we know in part, and we prophesy in part. But when that which is perfect is come, then that which is in part shall be done away." and the 12th verse: "For now we see through a glass darkly; but then face to face: now I know in part; but then shall I know even as also I am known."

Some teach that the perfect in this verse means the completion of the New Testament; but if that were true then we would have completeness in knowledge; we would know God even as He knows us; and we would be face to face with Jesus, because the New Testament was completed in 100 A. D. This is hardly the truth. The "perfect" cannot apply to the competition of the New Testament because none of these promises have come true! The "Perfect" is Jesus Christ, when He comes we will know as God knows us and we will be face to face.

I don't believe any Christian minister, evangelist, missionary, Bible teacher, or any Christian would dare say, that now that we have the New Testament: "we know God as He know us." or, that the New Testament brings us: "face to face with Jesus". As much as the New Testament illuminates our knowledge and experience in Christ, we still are looking somewhat, through the glass darkly until we are in His presence.

If the 13th chapter of 1st Corinthians is saying the Gifts of the Holy Spirit stopped at the completion of the New Testament; then why does the 14th chapter start with the instruction: "Follow after love, and desire spiritual Gifts, and that you may prophesy."

In the original it would read: "Eagerly pursue and seek to acquire love; make it your aim, your quest; and earnestly desire and cultivate the spiritual endowments especially that you may prophesy; that is, interpret the divine will and purpose in inspired preaching and teaching." How could the Gifts be stopped when the Bible asks us to covet and eagerly pursue spiritual Gifts?

Just for the readers information there are three complete chapters in the New Testament devoted completely to the teaching about the Gifts of the Holy Spirit and their operation. They are found in 1st Corinthians chapters 12, 13, and 14.

To write the New Testament the disciples, apostles, and members of the early church gathered and documented all the material that went into its completion. They depended upon the leadership and guidance of the Holy Spirit to direct them in each word they would use in writing this Word of God. They were so committed, so dedicated, so emptied of self, so inspired of God, that they were considered to be "Holy Men" in God's sight!

The Bible was not thrown together by a bunch of misfits! No group of do-gooders decided it would be nice to have a book with some stories to stir our emotions to make us feel better; it is the everlasting inspired Word of the living God, and we better respect it, and give heed to it, because our eternal soul's future depends upon our relationship with His Word!

When the New Testament was canonized it contained exactly what God wanted to be in the New Testament! What God did not want in the New Testament, was not put in the New Testament. If God had intended the Gifts of the Spirit to stop at the completion of the New Testament; He would have omitted those three chapters along with many other verses of scriptures about the Gifts of the Holy Spirit, at that time. The very fact that they are in the New Testament proves that God wanted the Gifts to continue even for us today!

The Bible is clear about the Gifts of the Spirit. They are listed in 1st Corinthians 12: 8-10 and they are:

> The Gift of the Word of Wisdom
>
> The Gift of the Word of Knowledge
>
> The Gift of Faith
>
> The Gift of Healing
>
> The Gift of the Working of Miracles
>
> The Gift of Prophecy

The Gift of the Discerning of spirits

The Gift of Speaking in Tongues

The Gift of the Interpretations of Tongues

The Gifts of the Spirit are not to be confused with the Fruits of the Spirit, which is listed in the Book of Galatians 5: 22, 23; and they are:

The Fruit of Love

The Fruit of Joy

The Fruit of Peace

The Fruit of Longsuffering (patience)

The fruit of Gentleness

The Fruit of Goodness

The Fruit of Faith

The Fruit of Meekness

The Fruit of Temperance

Very seldom will one hear a message given on the Holy Spirit; and probably never on the subject of the Gifts of the Holy Spirit. It might be upsetting to the congregation. However, to be acceptable; many ministers have presented the Fruits of the Spirit as the Gifts of the Spirit.

I once heard a minister say that the running ability of Emmitt Smith, a very successful running back for the Dallas Cowboys, was a gift of the Holy Spirit. A former president of the Moody Bible Institute has a DVD titled: The Gifts of the Holy Spirit" in which he explains our ability, as Christians, to show compassion, was a gift of the Holy Spirit. Note: Compassion is a human attribute which every person has, and has the ability to express. The Holy Spirit may move us with compassion to help someone in need or lead us in making a decision; but compassion is not listed as a Gift of the Holy Spirit.

A noted television minister considers Mercy as a gift of the Holy Spirit and says so in a book he wrote. The Gifts of the Spirit are for a special purpose in the ministry of the body of Christ. The Fruits of the Spirit are those human characteristic that we all have; the difference is; in the Christian these emotions are steadfast in the spiritual realm. As we are willing, the Holy Spirit will take us and begin to make us into the likeness of Christ. That is part of His office.

There are three members of the Godhead family; God the Father, God the Son, Jesus Christ, and God, the Holy Spirit. They are spiritual; but can take on physical form and each has a distinct purpose and ministry. Because of this, the Christians in India, Africa, Europe, and South America, can experience the same preciousness and nearness of God, at the same time, as we do here with Christ in America. They can work individually with the individuals or collectively with a group. They are all powerful in their purpose; yet are pleased to let us minister this power according to their will. This they do through the Gifts of the Holy Spirit that is given to us..., and they, the Godhead are pleased to respond to our prayers.

Lets consider the gift that is most controversial; the awful, dreaded Gift of tongues. I remember the first time I came in contact with this strange repulsive thing. A young man by the name of Steve Hill was holding meetings in a school building that I attended. After several weeks of wonderful Bible study, he said some words about speaking in the Spirit and I thought to myself; Don't tell me this handsome young man is involved in that! So I know that tongues are a turn off subject. However, I must say it's kind of neat, to pray to God in Chinese.

First, let me be clear; the Gift comes from God, not Satan. God is not going to give you something that is bad! Its purpose is to edify the believer. In 1st Corinthians 14: 4 we read: "He who speaks in tongues edifies and improves himself." The believer's faith is increased because he is speaking to God with praise that is coming from his inner most being. It is a prayer language that by passes our understanding; but is understood by God; for it is the Holy Spirit praying through us. Romans 8: 26, 27:

> "Likewise the Spirit also helps our weaknesses; for we know not what we should pray for as we ought: but the Spirit itself makes intercession for us with earnest compassion, groaning, which cannot be uttered. And he that searches the hearts, knows what is the mind of the Spirit, because He makes intercession for the saints according to the will of God."

When we are praying in the Spirit; the Holy Spirit is praying through us according to the will of God, and He is also interceding on our behalf. That is why verse 28 is true! If we are communicating with God through praying in the Spirit, we are praying according to the will of God, and the Spirit is interceding for us, and then we can say with assurance:

> "All things work together for good to them that love God, to them who are called according to His purpose."

How many times have you heard people quote this verse out of context saying: "All things work together for good", and apply it to some every day occurrence. It is intended for the Christian who is walking, living, and praying in the Spirit.

In group worship, praising God in the Spirit magnifies and glorifies God. In Acts 10: 44-48:

THE CONSCIOUSNESS OF THE HOLY SPIRIT

> "While Peter yet spoke these words, the Holy Spirit fell on all them which heard the word. And they of the circumcision (the Jews), which believed were astonished, as many as came with Peter, because that on the gentiles also was poured out the gift of the Holy Spirit, for they heard them speak with tongues and magnify God. Then answered Peter, can any man forbid water, that these should not be baptized, which have received the Holy Spirit as well as we?"

There are several things we can learn from these passages of scripture. Peter with some disciples were invited to the Gentile home of Cornelius to minister to him and a group of friends about Jesus Christ. And while Peter is speaking, the gentiles opened their hearts to Christ and were all filled with the Holy Spirit and began to praise and magnify God with tongues, in the Spirit. This was another Pentecost because Peter said: "they received the Holy Spirit as well as we" or in the same manner as we did! Then it states that they heard "them speak with tongues and magnify God." The use of tongues in praying, speaking, and singing in the Spirit; glorifies and magnifies our heavenly Father.

In Ephesians 5: 18, 19 it says:

> "And be not drunk with wine, wherein is excess; but be filled with the Spirit; speaking to yourselves in psalms and hymns and spiritual songs, singing and making melody in your heart to the Lord; giving thanks always for all things unto God and the Father in the name of our Lord Jesus Christ."

And in Colossians 3: 15, 16 we read:

"And let the peace of God rule in your hearts, to the which also you are called in one body; and be thankful. Let the Word of Christ dwell in you richly in all wisdom; teaching and admonishing one another in psalms and hymns and spiritual songs, singing with grace in your hearts to the Lord."

God loves the praises of His people!

Notre Dame University held a huge charismatic meeting one evening and I had the privilege of attending. We arrived late, it was getting dark, and we didn't know were the meeting was to be held. Two students were walking along the sidewalk and we pulled up beside to ask where the meeting might be. They said; listen, just follow the sound of the music. And then we became aware of the sound of this music floating across the campus. It was nothing like I had ever heard before. It was coming from the football stadium and we drove over there.

We climbed the stairs and entered the stadium and realized some fifty thousand Christians were there singing in the Spirit. That's in tongues.

People were singing in different keys and on different notes, which should have been a total mess. The judges on American Idol would be holding their ears; but this was not the sound of people vocalizing on different notes and pitches; this sound; only heaven could produce. The voices blended with breath taking harmony the volume rose and lowered in complete unison; it was like angels singing. They were singing in the Spirit!

I might add what the scene looked like. There was a stage at one end of the field were five or more archbishops of the Catholic Church sat. They were dressed in white robes with high white hats; I think, as I remember, one may have been red. There must have been two to three hundred priests, all dressed in white robes, standing in the circle of the playing field of the stadium. There were breaks in the program and during that time the priest would blow kisses up to the crowd and the crowd would return them.

We were all served communion, and when it was my turn to receive, I commented to the priest that he had just served communion to a Baptist and the priest said; Praise the Lord!

Rain had been predicted and as we were facing west, we could see the dark cloud coming. From the platform came the request for prayer to have the Lord stop the rain and as we watched the approaching cloud parted and passed around the stadium. That was a meeting that surely pleased God immensely. If as a Christian, you have never experienced singing in the Spirit; then you have something special to look forward to, and I hope it's before you get to heaven.

Paul states he speaks in tongues more than all the others and adds:

> "Yet in the church I had rather speak five words with my understanding, that by my voice I might teach others also, than ten thousand words in an unknown tongue."

This is clear; Paul would rather speak in understandable words than in tongues, or an unknown language. Then when and where did he speak in tongues more than them all? The answer is that Paul used the Gift in his personal devotion time. I have had the Gifts of the Holy Spirit for some forty years; but I have never spoken in public in tongues, or another language. It has always been in my devotions and private life. Now, I have sung in the Spirit in worship services, but not as an individual in public.

One of the Gifts is the interpretation of tongues; I have had an experience with this on one occasion. During our meetings, one of the women of our church always had something from the Lord to share and in almost every meeting she was up with a prophecy, an interpretation, or a special message. We men wished she would shut up. Well, at this time, the movement of the spirit filled life was very new in our churches and the community in which we lived. Our group had about six months experience with the baptism of the Holy Spirit and so often we would be asked to send someone to their church to teach and explain about the Gifs of the Spirit.

This woman and I were selected to go to this Episcopal Church to teach them and during a quiet time she stands up with tongues and I said to my self why doesn't she keep quiet; and just then, in my spirit I had the interpretation. God has a sense of humor. This was something new to me, but I was conscience to the fact that God wanted me to say something, so I got up and began to speak. I spoke about the history of the Episcopal Church, which I knew nothing about, and I went on for several minutes and closed with; this is the word of the Lord. Evidently it was accurate because there was a good response from their priests when the meeting was over. What we should learn, is the Holy Spirit does not throw you around on the ground, out of control, or behave in an unseemly manner, I was in control the whole time.

I have not had an interpretation of tongues since then; but that could be because we moved to another state, in regards to my work, and the church we worshiped in did not believe in the Gifts.

By reading the 14th chapter of 1st Corinthians one will learn that in the Church, teaching is more important than tongues unless there is an interpretation; then they are equal in value. There are strict rules in regards to the operation of the Gifts of the Holy Spirit in the church. One

THE CONSCIOUSNESS OF THE HOLY SPIRIT

cannot give a message in tongues, unless there is someone present who can give the interpretation of the tongues. If a person gives a message in tongues and there is no interpretation, then it is known by everyone that the person that spoke did not speak by the leading of the Holy Spirit. In addition, if someone gives a message in tongues, and there is an interpretation; there must also be the confirmation by two or three witnesses that the message was from God. This way, God has built into the system a sure way to keep out someone who would like to show off.

1st Corinthians 14: 28 it states: "But if there be no interpreter, let him be silent in the church." At no time, does the Holy Spirit over rule or cause disorder or confusion in the operation of the Gifts. If this takes place, then everyone will know this activity is not of the Holy Spirit. The Holy Spirit is subject to the spirit of the Christian as to speak or not to speak.

In Lester Sumrall's book: "The Gifts and Ministries of the Holy Spirit"; he gives the account of a meeting in Washington, D.C. where a brother gave a message, which another person interpreted. Afterwards, a young man came forward to speak with the man who had given the message; but he spoke in a foreign language. The brother said, "I'm sorry, but I don't understand you." The young man replied, "But you spoke to me in my language, I am Persian." The brother answered, No, I do not know your language; that was God speaking to you through the Holy Spirit. This led the young man to accept Christ as his Savior.

All of the Gifts of the Spirit were at work in the early church as recorded in the Book of Acts. They are still working today in churches and mission fields where they are believed. Some of the Gifts can be two fold such as pointed out in Lester Sumrall's book the case when Peter said: "And the promise is unto you, and to your children, and to all that are afar off, even as many as the Lord our God shall call." How did he know this? It could have been a word of Knowledge,

Wisdom, or Prophecy; does it make a difference; no, whatever it was, it was a revelation of God's truth. When we see Peter we can ask him which it was.

As above, Faith, Miracles, and Healings could go together. Did faith bring about the healing or was the healing a miracle. I know of a church that was in the process of renting some property that would improve its ministry, and they didn't know if the deal would go through or not and so after a meeting they met for prayer and in prayer a member had a word of knowledge that the deal would go through, so then all went home and got a good nights sleep. The next day sure enough the deal did go through.

Healing is another Gift of the Holy Spirit and the Book of Acts has many cases of healings that the Holy Spirit did through the disciples and early Christians. I myself have witnessed close to a hundred instant healings that have been remarkable. When the teaching about the Holy Spirit first came to our area, it was very new. Most of us didn't believe in the Gifts because we grew up in the Baptist church. Different members were receiving the baptism of the Holy Spirit and praying in the Spirit, and we didn't understand what was going on. Then a couple visited a meeting in Chicago, where Kathryn Kuhlman was holding meetings and they came back telling us they had seen people healed. We couldn't believe it!

The workers with the Jesus Film ministry tell us one way for sure they can get into a closed area is that they can assure the head of the village or city that they will bring physical healing when they come. In John 14: 12 Jesus said to His disciples:

> "Truly, truly, I say unto you, He that believes on me, the works that I do shall he do also; and greater works than these shall he do; because I go unto my Father."

THE CONSCIOUSNESS OF THE HOLY SPIRIT

How does one receive the baptism of the Holy Spirit? If you are a Christian, you receive the Holy Spirit the same way you receive Jesus Christ as your Savior. You simply pray and ask God to baptize you with the Holy Spirit and He will.

If you are not a Christian, you can be one in just one minute by receiving Jesus Christ as your Savior. If you believe you are a sinner and you want God to forgive you, just pray to God and tell Him you are a sinner and ask for His forgiveness. Thank Him for sending Christ to die for you on the cross and tell God that right now; you are receiving Jesus Christ as your personal Savior. If you mean this from your heart, you will know that something happened to you on the inside.

This decision is based on the Word of God. John 3: 16:

> For God so loved the world that He gave His only begotten Son, that whosoever believe in Him, shall not perish, but have everlasting life."

From this verse we learn that God loves us, that Christ died on the cross in our place so that we can be forgiven and have everlasting life. You accept the truth of that verse, as a promise from God that you are His Child! Then the next step is to tell somebody what you did.

Romans 10: 9, 10: "That if you confess with your mouth that Jesus is Lord and believe in your heart that God raised Him from the dead, you are saved. For with your heart you believe unto righteousness and with the mouth confession is made unto Salvation." There is no religious ritual, no special ceremony or joining a church; it is just you and God and Jesus Christ.

The atheist boasts that faith has no substance to support it; it is a belief system that cannot be logically justified; but they are the ones scraping everywhere to find something, anything possible to support their theory. The Christians belief system is founded upon a Book that represents 6000 years of history and is supported by evidence from science and the many millions of people, who down through the ages, have accepted and believed the Biblical record.

The Gift of tongues: praying in the Spirit and singing in the Spirit, is a personal Gift that I am blessed with and enjoy each day. The Holy Spirit can work in amazing ways. On a Monday morning, some years ago, the captain of the Wheaton College cross country team came to the platform to confess that he had broken the school's rule and dishonored God by taking the team into Chicago to run in a tri-state college competition. They had won but it was wrong, and he was sorry and asked for the school's forgiveness.

After a pause, Dr. Edman rose and came to the podium and asked if someone else had a confession they would like to make. Another student came to the platform to confess he had not been living like he should, and that he had sin in his life. Another came, and then another and soon there was a line of students waiting for their turn to come to the platform to make their confession. Some stood for two hours until it was their turn to speak. This went on all that day, all that night, all the next day, and all that night, until Friday morning at chapel. This was the result of prayer by the student body. God had laid it upon the hearts of six freshmen to pray each evening in the basement of the chapel for God to send revival, and He did.

What is the need in your life? Whatever it is, God will meet that need if you will turn it over to Him!

❋ ❋ ❋

CHAPTER VI

THE CONSCIOUSNESS OF THE TIME

On the roof of an ancient building is the form of a man pacing back and forth. He stops and looks out into space for a minute and then continues. He has been there all night and as the sun is starting to rise above the mountains, we realize he is mumbling something to himself. As we draw near we begin to understand the words he is saying: "Oh, if I only had the wings of a dove, I would fly away!" Psalms 55: 6

The man is King David and just that night, Nathan the prophet had told him of a rich man who had taken the one little lamb of a poor man and killed it for his own dinner. David was angry and shouted; who is this man! He shall be killed. Then Nathan raised his hand and pointed his finger at David and declared: "You are the man!"

Perhaps you know the story. King David had taken the wife of another man, Uriah, and committed adultery. They tried to cover the act by having her husband come home from the war and when David learned that Uriah did not sleep with his wife; he returned him to the front of the battle so he would be killed. II Samuel 12: 1-7

In our life time we face many temptations; many of us fall and have said to ourselves, like King David, Oh, if I could just get away from it all! There is a way; but it hurts. You have to admit to God that you are a sinner and receive Jesus Christ as your Savior. The God of heaven understands your feelings, hurts, and disappointments and He invites you to come to Him. In Isaiah 1: 18 God says:

> "Come now, and let us reason together says the Lord: though your sins be as scarlet, they shall be as white as snow, though they be red like crimson, they shall be as wool."

God established for us a blood stained cross that Jesus died upon to pay for our sins, and because every sin that you and I have committed were charged to Jesus Christ, we can have forgiveness.

The scripture states:

> "He was despised and rejected by men; He was smitten by God; He was wounded for our transgressions; He was bruised for our iniquities and our peace was upon Him; and by His strips we are healed. That all we like sheep have gone astray, we have turned everyone to his own way,

THE CONSCIOUSNESS OF THE TIME

> and the Lord has laid on Him the iniquity
> of us all."

In watching "The Passion of the Christ", we felt the horror of what Christ suffered in that crucifixion; but the film could not show what really took place! In Isaiah 53: 10, we read:

> "Yet it pleased the Lord to bruise Him; He
> put Christ to grief, when he made His soul
> an offering for sin."

Why would God be pleased in the suffering of His Son? Verse 11 continues:

> "He shall see the travail of His soul, and
> shall be satisfied: by His knowledge shall
> my righteous servant justify many; for He
> shall bear their iniquities."

These are not words from the New Testament; these are words spoken in prophecy some seven hundred years before Christ was born. In the twelfth verse of Isaiah 53; it speaks of Christ pouring out His soul unto death in securing our Salvation.

Jesus came into the world as a baby to grow up fulfilling all of the Old Testament requirements of perfection in order that He could become the perfect Lamb of God. The Jewish nation rejected Him and crucified Him and in so doing brought God's judgment upon them. Jesus prophesied in Matthew 24: 2 that this judgment would take place, that Jerusalem would be destroyed and the Jewish people would be dispersed throughout the world.

This took place in 70 A D., and as we know, later, they experienced persecution above that of any other nation. During WW II, six million were killed in the most inhuman manner. God did not look kindly upon those who reject His Son. But today we are living under the time of forgiveness; when all that we need to do is to acknowledge Jesus Christ as our Savior and Lord, the One who died in our place on the cross and we can have forgiveness.

If a person took that step, you know, to really get everything settled in one's life with God, and maybe the family too, how will God respond to that person? We have the answer in Matthew 11: 28, 29; where it says:

> "Come unto me, all you that labor and are heavy laden, and I will give you rest. Take my yoke upon you, and learn of me; for I am meek and lowly in heart; and you shall find rest unto your soul. For my yoke is easy and my burden is light."

God is saying here if you have burdens, troubles that are weighing you down, to bring them to Him. He will give you rest, and inner peace of mind.

Even the most disgraceful sins and crimes can be forgiven and the guilty made clean to start a new life again. In the Book of Colossians the second chapter, verses 13 and 14 we read:

> "And you, being dead in your sins and the uncircumcision of your flesh, has He quickened together with Him, having forgiven you all trespasses; blotting out the handwriting of the ordinances that was against us, which was contrary to us, and took it out of the way, nailing it to His cross.

What is being said here are that the records of all our sins will be removed, deleted, because they were paid for, nailed

THE CONSCIOUSNESS OF THE TIME

to the cross of Jesus Christ. That's why the Christian can say:

> "I am crucified with Christ, nevertheless I live; yet not I, but Christ lives in me; and the life which I now live in the flesh I live by the faith of the Son of God, who loved me, and gave Himself for me." Galatians 2: 20

In these chapters we have tried to present enough evidence to at least, cause the reader to consider that the Bible is not a book of myths; but that it stands in the unique position of having recorded the complete history of the universe, with all that is within it, and reveals the heart and mind of God as our creator.

There are many different levels of faith that we express each day that are based on the decisions we make; but the decisions are not made on blind faith. We believe the car will start and when it doesn't we are surprised! The evidence is that there sits a huge piece of machinery that has started consistently for two years and now it won't. With a little analysis we conclude the battery is weak and needs to be replaced, otherwise the car is good.

We should give the same consideration for the Word of God! Its message has been around for 6000 years and if in your case it hasn't rang a bell; reconsider the information in this book; or best, start reading the Gospels in the New Testament.!

I would like to speak personally to the reader. I'm 89 years of age and served in WW II. As a child we had no electricity, we went to church in a horse and buggy, there was no running water in the house, no radio. In 1928 we moved to Chicago. America was still coming out of the depression years and I still remember standing in line with my father to receive a quart of milk and a stick of butter every other week or so. I have watched America as she grew.

We have come a long way, but false teachings and false leaders have slowly removed the basic Christian foundation this country was founded upon. I fear America's greatness lays in it's past. I believe we are at that point in our history that unless the moral decay within us is stopped, and America turns back to God, we will not weather the attacks from our enemies. The violence that was once on distant shores has come to us. We neglect and refuse to recognize our enemies, even though they have loudly and openly defiled us. Here is a quote that should wake America up; but will it?

> "The process of settlement (of Islam in the United States) is a "Civilization-Jihadist" process with all the word means. The Ikhwan must understand that all their work in America is a kind of grand Jihad in eliminating and destroying the Western civilization from within and sabotaging their miserable house by their hands and the hands of the believes so that it is eliminated and God's religion is made victorious over all religions. with out this level of understanding, we are not up to this challenge and have not prepared ourselves for Jihad yet. Wherever he lands until the final hour comes, and there is no escape from that destiny except for those who choose to slack"

The word: "Ikhwan" is the Arabic word for "brothers" and refers to the Islamic religious militia dating back to the Arabian ruler Ibn Saud in 1926. Notice the words: "destroying the Western civilization from within". These words are not found in some secret book hidden away out of sight;

but published in our newspapers in America and on the internet under: "The Muslim Brotherhood". (Ikhwanweb)

In Matthew 24: 37, Jesus is speaking about His return and states:

> "But as in the days of Noah were, so shall also the coming of the Son of man be. For as it was in the days before the flood they were eating and drinking, marrying and giving in marriage, until the day that Noah entered the ark, and knew not until the flood came, and took them all away; so shall also the coming of the Son of man be."

The normal social activities spoken here were not what Jesus is speaking about; He was referring to corrupt activities, the lack of morals within the society. This is confirmed in Genesis 6: 11-13:

> "The earth was also corrupt before God, and the earth was filled with violence. And God looked upon the earth, and behold, it was corrupt; for all flesh had corrupted His way upon the earth."

I believe we are seeing for the first time, the earth being filled with violence. Jesus said this is the way the world will be when He returns; a corrupt world full of violence. A society becomes corrupt as it removes the influences of God's laws and His standard of living. America has removed the Ten Commandments as the document to base our laws upon and replaced it with a humanistic philosophy. We have removed the Bible and prayer from our schools and replaced them with teen pregnancy, drugs, and crime.

Without clear Biblical understanding of God's standard; we do not know the dangers of pornography and sex video games on the internet. We have reached the point where every kind of evil is ok; but by all means, let's keep God out of the picture, and: "We don't want the divine foot in the door".

The wars we have known before have been restricted to certain countries; Europe, Vietnam, Korea and now Iraq and Afghanistan; but now we have Muslim terrorists that are everywhere, and even within our government and military.

Our state of mind is to report anything looking suspicious, even the shoes grandma maybe wearing at the airport, have to be inspected. The opposition to Christian principles has nothing to offer society that inspires and uplifts the individual to want to be a better person. It has no code of conduct by which to live, or any hope beyond the grave. We can observe the difference between our Christian America and the other countries with their religions. The orient has Buddhism, Shintoism, and Confucianism. India has Hinduism and the Muslim lands have Islam. Which other country would you rather live in, than America?

In considering those other religions we should note: Japan has between 100,000 to 120,000 Muslims living in their country; 90% of these are of foreign origin and only 10% ethnic. Does this seem strange to you?

England has 818,000 Muslims living there. France has six million Muslims; Italy 1,210,000 Muslims with only 50,000 having Italian citizenship. Germany has two million Muslims and in Indonesia 86 % of the population is Muslim. Here in the United States we have close to seven million Muslims. (Wikipedia, the free encyclopedia)

All of the bombing attacks of the past, here in America and abroad, have been committed by Muslims. Yet the Muslim community in America is silent; no comments about the unjust killings by their people, not even when it's Muslims killing Muslims, there is no reply.

THE CONSCIOUSNESS OF THE TIME

In September 25, 2008, Geert Wilders, Chairman, Party for Freedom, the Netherlands, spoke at the Four Seasons Hotel in New York about the concerns the Muslims are bringing to the countries in which they are living. He writes: "There is a tremendous danger looming, and it is very difficult to be optimistic. We might be in the final stages of the Islamization of Europe. This not only is a clear and present danger to the future of Europe itself, it is a threat to America and the sheer survival of the West. All throughout Europe a new reality is rising: entire Muslim neighborhoods where very few indigenous people reside or are even seen. They become Muslim ghettos controlled by religious fanatics."

The Pew Research Center reported that half of French Muslims see their loyalty to Islam as greater than their loyalty to France.

On November 5, 2009, Maj. Malik Hasan, a Muslim at Fort Hood, shot and killed 12 military personal and wounded 31 others. On December 25, 2009, Abdulallab, a Muslim man tried to blow up a transatlantic flight on its way to Detroit and on May 4, 2010, Faisal Shahzad, a Muslim, failed to set off a bomb in New York City.

In contrast, consider the actions of our Pentagon who disinvited Franklin Graham, Billy Graham's son who was asked to speak there on the National Day of Prayer and who is the honoring co-chair of that organization; not to come because he made a statement on 9/11/2001, that the Islam religion was evil.

Could the world wide violence of Muslim terrorism be that which Jesus spoke of when describing the conditions of the world when He comes? When He comes this time, there will be no announcement of His coming, because it will take place in the twinkling of an eye. I Corinthians 15: 51, 52:

> "Behold I show you a mystery; we shall not all sleep, but we shall all be changed, in a

moment, in the twinkling of an eye, at the last trump: for the trumpet shall sound and the dead shall be raised incorruptible, and we shall be changed."

We are told in Luke 21: 34:

"Take heed to yourselves, lest at any time your hearts be overcharged with careless living so that day comes upon you unawares."

Romans 13: 11- 13 warns:

"And that knowing the time, that now it is high time to awake out of sleep: for now is our salvation nearer than when we believed. The night is far spent, the day is at hand: let us therefore cast off the works of darkness, and let us put on the armor of light."

In First Thessalonians 5: 6 we are warned:

"Therefore let us not sleep, as do others; but let us watch and be sober."

In 1ST Peter 4: 7 we read:

"But the end of all things is at hand: be you therefore of sound mind, and watch unto prayer."

Many books have been written on the subject of prophecy and most of the prophecies have been fulfilled, or are being fulfilled, even while I am writing this book. The next prophecy to watch is the attack on Israel by the surrounding nations, which could come at any time.

THE CONSCIOUSNESS OF THE TIME

In closing this book, let me be open with you; and you be open with me! Way back when I was in school I was assigned to write an article on Criminology. One of the books I used in my research was a text book used by the University Of Chicago Law School. In it was this example.

A tramp is walking along the road and in passing this house he sees this little blond headed girl playing in the driveway with a few coppers. He grabs the girl and kills her, throws her body in the ditch and runs off with the coppers.

The book asks the question: Should this man be found guilty for what he did?

The answer was: No, because he was just responding to his natural brute beast instincts!

Evolution offers no hope or reason to live. It gives no purpose of life. If there is no God we can live anyway we wish and that is exactly what we see in the world and in America today!

Our young people are being deceived into believing that truth, morals, and responsible living are not relative. All their friends are doing it; why are our parents so restrictive. Where do these ideas come from? They come from their schools.

A mother going through some of her son's things in his room found condoms and in confronting him learned they were dispersed by his grade school. Liberal thinking would say: "What's wrong with that?" What's wrong is that the tramp killed the blond haired girl and .....it was not so bad!

One thing America and the world can be sure of; a judgment day is coming. If you live out your life, and you miss heaven, then you have missed everything!

AND EVOLUTION IS A MYTH!

Philippians 1: 20

❅ ❅ ❅

DOCUMENTATION

CHAPTER 1

1. Speech given by Dr. Francis Collins, former director of the National Genome Research Institute and director of National Institute of Health.

2. The Ninety and Nine sung by Elizabeth F. Cleplane and Ira D. Sankey. 1874

3. "The Love of God" by Frederick M. Lehman Mays and arr. by Claudia Leman Mays.

CHAPTER II

1. Archaeological Supplement by G. Frederick Owen D.D., Ed.

2. Ibid: King Sargon, page 342.

3. Ibid: The Tell el-Armarna tablets. Page 312

4. Ibid: page 313.

5. Ibid: The City of Dothan. Page 309.

6. G. Frederick Owens's statement, page 309

7. The Black Obelisk, page 321.

8. Ibid: The Dead Sea Scrolls. Page 325.

9. Absalom's Pillar, page 311

ANTHROPOLOGY

10. Refuting Evolution, by Jonathan Sarfati, Ph.D. Page 16.

11. Ibid: Pages 16-17.

12. Wikipedia, the free encyclopedia, The Nebraska Man.

13. Ibid: The Java Man.

14. Ibid: The Piltdown Man.

15. National Geographic Magazine, October, 2008. Neanderthal Man.

16. Wikipedia, the free encyclopedia, Neanderthal.

17-19. Lucy, Wikipedia, the free encyclopedia.

20. Oldest Skelton of Human Ancestor Found. National Geographic Magazine, web page of Ardi.
Wikipedia, the free encyclopedia, for Ardi.

21. Refuting Evolution. Page 47.

22-26. Ibid: pages 16, 17, 18, 47, and 48.

DOCUMENTATION

27. Archaeological Supplement, page 363, the flood an "Ur".

28. CBS News, 60 Minutes, November 15, 2009 and December 26, 2010: The finding of B Rex.

29. Shattering the Myth of Darwin, by Richard Milton, AIG.

30-31. Mt. St. Helen. Wikipedia, the free encyclopedia.

32. Potassium-argon dating; A.A. Snelling; Refuting Evolution, AIG .pages 110-112.

33-34. Ibid: pages 111-112

35. Guy Ottewell's 1000 Yard Model.

36. NASA at World Book.

37. Refuting Evolution, page 121.

38-39. M. Denton, Evolution: A Theory in Crisis. Page 124

40. Jerry R. Berman's Book: The Origin of Life,

41. Not By Chance, by L. Spetner in Refuting Evolution by Jonathan Sarfati. Page 126.

42. Newsweek, March 23, 2007, "Beyond Bones and Stones, by Sharon Begley.

43. Refuting Evolution. Page 121.

44. Ibid. 124-126

45. Ibid pages 20-21 E.D. Wilson; The Humanist.

❋ ❋ ❋